Atoms in Molecules

An Introduction

Atoms in Molecules

An Introduction

Paul Popelier
UMIST, Manchester, UK

PRENTICE HALL

An imprint of PEARSON EDUCATION
Harlow, England · London · New York · Reading, Massachusetts · San Francisco · Toronto · Don Mills, Ontario · Sydney
Tokyo · Singapore · Hong Kong · Seoul · Taipei · Cape Town · Madrid · Mexico City · Amsterdam · Munich · Paris · Milan

Pearson Education Limited
Edinburgh Gate, Harlow
Essex CM20 2JE
England

and Associated Companies throughout the World

First published 2000

ISBN 0 582 36798 0

British Library Cataloguing-in-Publication Data
A catalogue record for this book is available from the British Library.

Library of Congress Cataloging-in-Publication Data
A catalog entry for this title is available from the Library of Congress.

Set by 35 in Times and Melior
Produced by Addison Wesley Longman Singapore (Pte) Ltd.,
Printed in Singapore

To Fred and Hermine
my favourite bond

To all researchers
Past, present and future
Contributing to AIM

Contents

List of box contents

Preface

Listen carefully to Nature for she only whispers

This book is an invitation. It is an invitation to participate in the quest to construct a solid bridge between modern computational quantum mechanics and chemistry. Nowadays, affordable computers run sophisticated quantum chemical programs yielding the energies and geometries of vast numbers of molecules. The development of chemical insight, however, has lagged behind these spectacular advances. Chemical concepts taught to undergraduates are based primarily on knowledge that was available in times when computers did not exist or were rare tools. Although these chemical concepts are undoubtedly valuable, they should constantly be confronted with modern solutions of the Schrödinger equation obtained by the present-day computers. Naïve reductionism set aside, we must admit that if the Schrödinger equation is true, then all of its consequences must be true. As a result, any chemical concept must be compatible with the Schrödinger equation; it cannot have a life of its own, disconnected from the physics that governs Nature. Of course this is a challenging research program, with many battles to fight, but I believe it is worth it.

Accepting that Science coerces us to a unified and streamlined picture of Nature, we face such questions as, how do we recover from quantum mechanics a chemically meaningful and intuitively appealing atom inside a molecule? How do we recover a bond in a molecule from quantum mechanics as a simple and computable object? How do we broadly classify bonds in terms of their covalent or ionic character based on quantum mechanics? How do we detect and characterize hydrogen bonding from modern wave functions? How do we *explain* the geometry of ligands around a central atom from the electron density?

Promising, rigorous and elegant answers to these and other important questions can be obtained from a theory called 'atoms in molecules' (AIM). This theory uses the electron density as its starting point. The electron density is on a par with the energy in the same way that eigenvalues and eigenfunctions go hand in hand. Nevertheless, the quantum chemical community has focused most of its attention on the energy and its derivatives and often describes the electron density as a digression or an appendix. Therefore a most thorough study of the electron density, such as AIM offers, automatically ends up as an esoteric curiosity. This situation cannot be justified, disregarding historical reasons, especially in view of the central place the electron density takes in density functional theory, which currently enjoys tremendous popularity. The electron density is a treasure of information and AIM shows how to reveal it. It should be stressed, however, that AIM is an *interpretative* theory, not an alternative computational scheme yielding the electron density.

The audience for this book is perhaps best described as 'to whom it may concern'. The decision to write the book was easy because I am convinced that AIM contains many nice ideas that are barely known. Deciding on the right level was much harder because I attempt to reach both the keen and mature undergraduate, as well as the experienced researcher. Anticipating criticism and possible frustration, I apologize in advance to both ends of the spectrum. The main goal of this book is to make AIM accessible in a fairly self-contained way to an audience wider than the one that reads the specialized research journals. I can imagine that this breath of fresh air can be useful to people outside computational chemistry such as organic and inorganic chemists, crystallographers, solid-state chemists, biochemists and physicists. This book is an invitation to anyone who is looking for a novel approach to chemical thinking. Two years of teaching AIM final-year undergraduates makes me confident that this book is appropriate for open minds. I have assumed only a basic knowledge of quantum mechanics and some exposure to computational chemistry.

Writing a comprehensive book about a novel paradigm is challenging. In a way, I have put new wine in new bottles, i.e. I have presented a large part of this novel theory called 'atoms in molecules' in a new way. Where appropriate I have included new data, examples and many figures in order to illustrate several concepts, but reiterating some old examples was unavoidable. The molecule methanal was chosen as the common thread throughout the book. Almost all chapters contain at least one 'box' in which I zoom in on the details of a particular concept. This allows readers to decide their own depth of understanding at any time. I have tried to avoid unnecessary mathematical detail but it is hard to please everyone. Also I make clear that this is *not* a review of AIM. Rather I have based myself mainly on Bader's monograph *Atoms in Molecules: A Quantum Theory* (Clarendon, 1990) with a desire to write a didactic textbook.

Writing a book about a novel paradigm as a textbook poses a dilemma in terms of referencing the literature. Originally I intended to provide no references at all except a brief reading list at the end of the book. Critical feedback on the manuscript made me insert a minimal list of references to the primary literature at the end of each chapter, as well as a more general 'further reading' list. Again I emphasize that it has not been my intention to provide an exhaustive reference list complementing Bader's monograph. Rather, I have inserted references only where the text may puzzle the reader. The theory of AIM is under constant development by several research groups worldwide. I apologize in advance to many authors who may find the reference list irritatingly incomplete and perhaps biased. This book is not a review but primarily a textbook attempting to explain AIM in a self-contained way, illustrated by examples from the Bader group and by a few own examples. Also, in view of the dynamic development of AIM, research-based controversies arise; I have deliberately avoided discussing issues which are not settled yet, although in places I have been tempted to speculate about their final outcome.

It is appropriate to reflect briefly on the way the scientific community perceives AIM. I have observed on many occasions that AIM strongly polarizes audiences in terms of sympathy. The reason for this phenomenon is multilayered and complex. It is important to ensure that AIM is not misinterpreted, misquoted or misjudged any more, owing to lack of knowledge. I would be very grateful if this book contributed to a fairer judgement of the theory. Admittedly, I have presented AIM with enthusiasm in the hope that it can be shared by others, but at the same time I give a balanced view of what it can and cannot do. In its current stage of development AIM does not provide all the chemical insight that a scientist would wish to have. There are many topics such as aromaticity, non-bonding interactions, the anomeric effect which are not properly linked to quantum mechanics. This is why an active research programme is being unfolded to realize the full impact that AIM can have in an amazing variety of areas.

Finally, it should be mentioned that the fairly slow acceptance of AIM has several causes. An important factor has undoubtedly been the computational cost of a topological AIM analysis. This factor dwindles almost on a monthly basis as the prices of computers are dropping faster than ever before. The second factor is simply that the theory requires one to sit down and come to grips with novel concepts which are entirely alien to an average chemist's theoretical background. This book has been written to address this communication or time investment problem. I hope that people will indeed take the time to absorb this new material and start using it to their benefit. A third factor is the lack of interest of the scientific community to 'fix something that ain't broke'. Could it be that generations of chemists have explained the right observations for the wrong reasons? Have they not used and abused orbital theory, for example, taken Lewis structures too literally or pushed the consequences of easily obtainable atomic charges to ridiculous limits? I am not claiming that AIM has all the answers, but I consider AIM as an invitation to reinterpret the physics of a molecule in terms of the chemistry that we have learned over the past centuries. In Chapter 1 I make a comparison between chemistry and the weather. If you are happy with that metaphor you will know what AIM is good for.

Paul Popelier
Manchester, 1998

Acknowledgements

I thank my parents for their support and the opportunity they have given me to become a more complete mind. Many thanks to Richard Bader who has initiated, led and guarded the development of the theory of 'atoms in molecules'. I am grateful to all of his co-workers who have painstakingly explored the foundations and applications of the theory, and to all researchers who have contributed to it. I owe my students Fiona Aicken and Sean O'Brien for their valuable feedback on the whole manuscript and their help with some calculations, to Ron Gillespie for his extensive comments, constructive criticism and support, Ian Bytheway for his thorough reading of Chapters 1–4, Richard Bone for feedback on Chapter 8, and Frank Mair. Many thanks to Alexandra Seabrook, Pearson Education's senior commissioning editor whose belief in this book has freed AIM from the oyster of specialization. I thank the independent reviewer for his considerable effort. Finally, I thank all readers who have accepted the invitation.

Permissions acknowledgements

Grateful acknowledgement is made for permission to reproduce material in this book previously published elsewhere. Every effort has been made to trace the correct copyright holders, but if any have been inadvertently overlooked the publisher will be pleased to make the necessary arrangement at the first opportunity.

We are grateful to the following for permission to reproduce copyright material: International Union of Crystallography for Figs 1.4 and 2.11; Addison-Wesley Publishing Company for Fig. 2.1; National Research Council Canada/NRC Research Press for Fig. 3.4; *Angewandte Chemie* for Fig. 3.5; *International Journal of Quantum Chemistry* for Fig. 3.6; Institute of Physics for Figs 4.4, 4.5, 6.6 and 6.7; American Chemical Society for Figs 4.2, 4.6, 10.4, 10.5 (Plate section), 10.6 and 10.7; Elsevier Science for Figs 4.8 and 9.5; Taylor & Francis for Fig. 9.6.

List of acronyms

3D	three dimensional
AIL	atomic interaction line
AIM	atoms in molecules
AO	atomic orbital
B3LYP	Becke–3–Lee–Yang–Parr
BCP	bond critical point
BP	bond path
CP	critical point
CCP	cage critical point
DCI	doubly excited configuration interaction
DFT	density functional theory
HF	Hartree–Fock
HOMO	highest occupied molecular orbital
IAS	interatomic surface
IR	infra red
LCAO	linear combination of atomic orbitals
LCP	ligand close packing
MG	molecular graph
MO	molecular orbital
MP2	second-order Møller–Plesset perturbation theory
(N)NA	(non)-nuclear attractor
NMR	nuclear magnetic resonance
RCP	ring critical point
SCF	self-consistent field
SDTCI	singly, doubly and triply excited configuration interaction
VSCC	valence shell charge concentration
VSEPR	valence shell electron pair repulsion

Software

New wave functions have been generated with the program GAUSSIAN (Gaussian 94, Revision B.1, M. J. Frisch, G. W. Trucks, H. B. Schlegel, P. M. W. Gill, B. G. Johnson, M. A. Robb, J. R. Cheeseman, T. Keith, G. A. Petersson, J. A. Montgomery, K. Raghavachari, M. A. Al-Laham, V. G. Zakrzewski, J. V. Ortiz, J. B. Foresman, J. Cioslowski, B. B. Stefanov, A. Nanayakkara, M. Challacombe, C. Y. Peng, P. Y. Ayala, W. Chen, M. W. Wong, J. L. Andres, E. S. Replogle, R. Gomperts, R. L. Martin, D. J. Fox, J. S. Binkley, D. J. Defrees, J. Baker, J. P. Stewart, M. Head-Gordon, C. Gonzalez, and J. A. Pople, Gaussian, Inc., Pittsburgh PA, 1995).

New AIM data have been obtained with the commercial computer program MORPHY98, which is available from http://www.ch.umist.ac.uk/morphy. Most figures were generated by MORPHY98, except for a few that can only be obtained from a future release version of MORPHY.

The original AIMPAC suite of programs, developed in Professor Bader's laboratory at McMaster University (Canada) can be freely obtained from http://www.chemistry.mcmaster.ca/aimpac.

Since 1994 the well known commercial program GAUSSIAN has provided an option to perform a restricted set of AIM utilities. Interested users should consult http://www.gaussian.com. However, in our experience the atomic integration subroutines *can* provide erroneous results, an uncritical interpretation of which may unjustly damage the reputation of AIM.

Chapter 1

The electron density

There will be another shock,
when we finally understand what it really means

1.1 The aim of AIM: a metaphor

Let us make a comparison between the weather and chemistry. Both encompass a complex world of phenomena which are ruled by a handful of physical laws. In the case of chemistry these physical laws are part of quantum mechanics, while in the case of the weather they are part of fluid dynamics and thermodynamics. Both quantum mechanics and fluid dynamics provide us with equations which are very hard to solve for real systems in spite of their appealing simplicity. The advent of sufficiently powerful computers has made possible predictions in both the world of chemistry and that of the weather. Solving the basic equations governing these systems requires elaborate and sophisticated computational schemes which typically yield millions of numbers. In turn, these numbers may be hard to interpret in terms of simple concepts such as, is this bond ionic or is it going to be foggy tomorrow?

How did chemists make predictions before the advent of computers? They gathered vast amounts of data on the phenomena they observed. Rationalizing these data led to rules which existed in their own right, disconnected from the physics of molecules. Strictly speaking, these rules could be applied only for the set of molecules for which they were originally observed. Similarly, at the dawn of meteorology, scientists perceived the weather as a complex reality imposing its own patterns and rules apparently independently of the physics that govern them. Parallel to this early scientific development, people (especially farmers) had been carefully observing the weather for centuries in an attempt to understand its behaviour. This knowledge, often expressed in colourful proverbs, could be vital, for example, to make a good decision when to harvest.

In summary, chemical concepts and explanations arise from observations at an 'upper' level, close to human experience and perception. In other words, chemical concepts emerged by looking at the world of chemistry itself, not the underlying world of physics. In a similar vein, weather types and proverbs arise from observations at this upper level, that of human experience, not by looking at lengthy solutions of complex differential equations describing the physics of the weather.

The main question is now: *is there a link between chemical concepts and quantum mechanics?* Or alternatively, keeping up with our metaphor, *is there a link between human weather awareness and fluid dynamics?* There must be a link because we live in only one world. Indeed, we cannot have chemistry existing all by itself, not being a product or consequence of quantum mechanics. Of course, the complexity of chemistry is expressed in novel phenomena that are not at all apparent from looking at a quantum mechanics textbook, but ultimately these phenomena follow from quantum mechanics.

Finding this link between the solutions of physics' equations and chemistry or the weather, however, is a major challenge. Some meteorologists may consider giving up their faith in the mathematical models and computational schemes of weather prediction. Some will be tempted to turn back to the wisdom of farmers and sailors, at least granting it the credibility of a useful complement to computation. Surely the situation is not that desperate in chemistry, but still the core of mainstream theories taught to undergraduates is based on old ideas rather than on *modern* solutions of quantum mechanical equations. These ideas date from times when quantum mechanics did not have the impact it has today and when computers were not that powerful or did not exist.

The theory of 'atoms in molecules' is a serious attempt to provide chemists with this as of yet incomplete bridge between quantum mechanics and chemistry. If we solve the Schrödinger equation for a molecule and obtain a 'telephone directory of numbers', how do we know which atoms are bonded? Can we define and study an atom *inside* a molecule? How does the bonding scheme change when nuclear positions are altered? How much charge is transferred from one atom to another? How much is the electron density polarized inside an atom? Where are the acidic regions in the molecule? Where is the preferred site for nucleophilic attack? For questions like that I invite you to have a look at AIM, the only theory that provides a simple, rigorous, useful and powerful definition of the two cornerstones of chemistry: the atom and the bond.

1.2 Quantum mechanics

There are many textbooks on quantum mechanics, and summarizing their content in this section would not only be impossible but also serve no useful purpose. Fortunately there is no need to do so in order to understand the rest of this book. We will review the necessary ideas and formulae as they are required, and for now it is sufficient to appreciate that most quantum mechanical knowledge has been implemented in a host of quantum chemistry computer programs that can be used by non-experts. We are aware of the dangers of such a black box approach, but users can consult appropriate documentation according to their background, agenda and degree of discomfort.

It is perhaps more useful to emphasize a few thoughts that are barely mentioned in theoretical chemistry texts, and in doing so render a human touch to quantum mechanics. It is fair to say that the very nature of quantum mechanics has divided people into two schools of thought. Dirac, Pauli, Heisenberg and Bohr belonged to one school, which summarized its view in what we now call the Copenhagen interpretation. The other school, encompassing Einstein, de Broglie, Schrödinger and Bohm, is not known by a named interpretation but has always stressed the physical clarity and common sense of any quantum mechanical interpretation. The proponents of this school have pushed and even tormented themselves and others to capture quantum reality in terms of concepts that can be grasped by human experience and intellect. Thanks to this attitude, Schrödinger was led to propose his most valuable equation, helped by his tenacity, mathematical skill and courage to argue against the scientific establishment. Indeed, his great intuition was guided entirely by the obsession to think of quantum phenomena as real, graspable objects. The Copenhagen school, on the other hand, seems to have surrendered to a quantum reality which is fundamentally bizarre and intrinsically beyond human comprehension. Bohr, the father of this school, was even prepared to give up certain philosophical principles and premises of the Science Enterprise, something which Einstein had always strongly resisted, being the strongest opponent of the Copenhagen school.

No matter how torn the pioneers of present-day quantum mechanics were, they left us with a consistent and extremely useful mathematical formalism to describe quantum phenomena accurately. This is probably the reason why the majority of scientists do not worry about the deeper physical meaning of this formalism. The most blatant example of this is that there is still no generally accepted explanation of the real meaning of the wave function ψ (Cramer, 1986), which is the solution of Schrödinger's equation. Schrödinger himself struggled with the significance of ψ, especially when it describes more than one particle. In this case, it becomes an object which resides in a space of dimension higher than three and therefore loses the direct 3D matter-wave interpretation that guided Schrödinger in 'guessing' his equation.

1.3 The electron density

It is gratifying to know that the wave function ψ determines all the information that can be known about a quantum system. Unfortunately, however, ψ cannot be directly observed. At least two problems arise if one wants to interpret and understand ψ directly for an arbitrary system of particles. One problem is that ψ is complex and the other is that ψ does not reside in real 3D space. We will discuss the solution to each problem in turn.

The first problem is overcome by introducing the product of ψ with its complex conjugate ψ^* or $\psi^*\psi$ (the complex conjugate of the complex number $a + bi$ is $a - bi$). The function $\psi^*\psi$ can also be written as $|\psi|^2$ which is the square of the modulus of the complex quantity ψ. This is not the only way to turn a complex quantity into a real one. This choice was most likely inspired by the analogy with classic wave theory where the intensity of wave motion, for example electromagnetic, pressure or sound, is proportional to the square of the amplitude. Schrödinger thought of individual electrons as wave packets which had a smooth and smeared-out 'charge density'. However, the physical interpretation Schrödinger originally had in mind for $|\psi|^2$ was replaced by Born's view which is now generally accepted.

Born, on the other hand, proposed that one finds the entire charge of an electron in some place with a probability proportional to $|\psi|^2$. So to quote Feynman's words: 'the electron is either here, or there, or somewhere else, but wherever it is, it is a point charge'. What were the grounds for Born's conviction of the correctness of associating wave amplitude with the number of particles? Ultimately, his interpretation goes back to the photoelectric effect, which actually – perhaps ironically – led to the Schrödinger equation via the matter-wave theory introduced by de Broglie. We will not reproduce here the details of Born's scattering calculations using analogies with optics, but it suffices to remind ourselves that in the photoelectric effect the *number of* incident *particles* (light quanta) is proportional to the *intensity* of the light (which is a wave phenomenon). The full and ultimate meaning of $|\psi|^2$ is still subject to active research building on work initiated by Dirac, Wheeler and Feynman many years ago, but the so-called transactional interpretation seems promising.

The second problem was that in general ψ is an object residing in a high-dimensional space. Indeed, each electron is described by four coordinates: three spatial coordinates denoted by $\mathbf{r}_k = (x_k, y_k, z_k)$ for electron k and one spin coordinate denoted σ_k (which can be either α spin or β spin). Hence, if there are N electrons in a molecule then its electronic wave function is described by $4N$ coordinates. How can this horribly complicated function be 'compressed' into a 3D function which can be more readily visualized?

This problem can be solved by integrating ψ over all the spatial coordinates except the set of three coordinates describing *one* electron, and by summing over all spin coordinates. The resulting function is called the *probability density* $P(\mathbf{r})$ or probability *per unit volume* of finding one electron regardless of its spin and the instantaneous positions of the remaining electrons. We can move on from the probability *density* to a probability by integrating $P(\mathbf{r})$ over a given volume. Consequently, the function $P(\mathbf{r})$ is mapped onto a number P. Suppose $P = 0.7$ for a given portion of space (say a box), then 7 times out of 10 we would find an electron inside this portion of space. This is an example of the literal statistical meaning due to Born as described above. As expected, integration of $P(\mathbf{r})$ over all space yields exactly one since the electron must always be somewhere in space. In other words, mathematical rigour requires that a probability distribution be normalized to unity.

If the probability density $P(\mathbf{r})$ is multiplied by the number of electrons N then we obtain the *electron density* or *electron distribution*, which is designated by $\rho(\mathbf{r})$. It immediately follows from the considerations above that $\rho(\mathbf{r})$ integrated over all space equals N and is therefore not a probability distribution, rigorously speaking. It is instructive, however, to think of $\rho(\mathbf{r})$ as a function yielding the *total* (accumulated) probability of finding *any* of the N electrons in an infinitesimally small volume of space. The function $\rho(\mathbf{r})$ is crucial to the development of the ideas in AIM and therefore we revisit $\rho(\mathbf{r})$ in more detail in Box 1.1.

Box 1.1 A closer look

The electron density from the wave function

Schrödinger's time-independent (or stationary state) equation is typically written as $\hat{H}\psi = E\psi$ where $\hat{H}$ represents the Hamiltonian operator and E is an energy solution and ψ a corresponding wave function solution. The Schrödinger equation is an eigenvalue equation which means that its solution is in fact a spectrum of eigenvalues (E) and corresponding eigenfunctions (ψ). For some unclear reason the quantum chemistry community has spent much more effort discussing and interpreting the energy (and its derivatives, fox example forces, force constants) than the wave function (or electron density ρ). This unequal treatment is untenable when one realizes that eigenvalues and eigenfunctions exist on a common footing. The theory of AIM restores this inequality by taking the electron density, which is derived directly from ψ (the eigenfunction), as its source of information.

For a fixed arrangement of nuclei, the probability of finding each one of N electrons in an infinitesimal volume element is given by

$$\psi^*(\mathbf{Q})\psi(\mathbf{Q})d\mathbf{q}_1 d\mathbf{q}_2 \ldots d\mathbf{q}_k \ldots d\mathbf{q}_N \quad (1)$$

where q_k are the set of four coordinates describing the k_{th} electron, i.e. the three spatial coordinated denoted by $\tau_k = (x_k, y_k, z_k)$ one spin coordinate denoted σ_k (which can be either α or β), so $\mathbf{q}_k = (x_k, y_k, z_k, \sigma_k)$. The symbol $\mathbf{Q}$ is shorthand for the total set of $4N$ coordinates, so $\mathbf{Q} = (\mathbf{q}_1, \mathbf{q}_2, \ldots, \mathbf{q}_N)$. Following this notation, the volume element can then be written as $d\tau_k = dx_k dy_k dz_k$. The physical meaning of the probability given above is the chance of finding a particular configuration in which each electron is in a particular position and has a given spin. We can now 'blur' this extremely detailed snapshot by summing over all spin coordinates and integrating over all coordinates except those of *one* electron. It does not matter which electron because all electrons are equivalent. For convenience we focus on electron 1 such that we integrate over the spatial coordinates of electron 2, and electron 3, and all electrons up to electron N. The probability of electron 1 being in the volume element $d\tau_1$, independent of the instantaneous positions or spins of the other electrons is then given by:

$$\sum_{\text{spins}} \left[\int d\tau_2 \int d\tau_3 \ldots \int d\tau_N \psi^*(\mathbf{Q})\psi(\mathbf{Q})\right] d\tau_1 \quad (2)$$

where $\sum_{\text{spins}}$ is the summation of both spins of all the electrons.

The total (accumulated) 'probability' to find *any* of the N electrons in $d\tau_1$ is obtained by multiplying equation (2) by N. This quantity corresponds to the total probability of finding electronic charge in $d\tau_1$. The maximum value this quantity can attain is N, which is why it is not a probability in the strict mathematical sense of the word (i.e. a true probability function must be normalized to unity). We can then turn this probability back into its corresponding probability *density*, that is the probability per unit volume, and obtain the familiar electron density ρ:

$$\rho(\mathbf{r}_1) = N\sum_{\text{spins}}\left[\int d\tau_2 \int d\tau_3 \ldots \int d\tau_N \psi^*(\mathbf{Q})\psi(\mathbf{Q})\right] \tag{3}$$

which can be abbreviated as follows, dropping the unnecessary index 1:

$$\rho(\mathbf{r}) = N\int d\tau' \psi^*(\mathbf{Q})\psi(\mathbf{Q}) \tag{4}$$

where $\int d\tau'$ means integration over the spin coordinates of all electrons and the spatial coordinates of all electrons but one. This mode of integration recurs throughout AIM.

1.4 The electron density from theoretical calculations

Again we are facing the hopeless task of summarizing a large part of quantum chemistry in this section, which is why we restrict ourselves to the key points and refer the reader to the many excellent textbooks and original papers for 'details'.

It is fair to state that, had it not been for the Schrödinger equation, quantum chemistry would not be nearly as developed as it is now. The only experimental values that appear in this equation are so-called fundamental constants such as Planck's constant h, the mass of the electron m_e and the fundamental charge e. A computational scheme that relies on no experimental values other than the fundamental constants is said to be *ab initio*. It is almost mysterious that so many molecular properties can be predicted by an equation which does not incorporate any experimental parameters other than the fundamental constants. Moreover if we write the Schrödinger equation in *atomic units* (details in Box 1.2) we need no experimental parameters at all.

The beauty of *ab initio* chemistry is its ability to predict chemical properties resorting only to computational schemes and computers. Of course, *ab initio* calculations can be very demanding which is why algorithms with severe shortcuts have been introduced in the past such as *semi-empirical* calculations. *Molecular mechanics* is a method which does not encompass any quantum mechanical formula at all, and which attempts to predict molecular properties based on external data, both experimental and theoretical. Such methods seriously depend on the quality of the experimental input but

Box 1.2 A closer look

Atomic units

There are good reasons to base our daily life units on atomic units, even if they do seem to be just another set of units to some sections of the global population that are still coming to terms with SI units. Indeed, the length we call one metre is still completely arbitrary although it is part of a convenient metric system. Atomic units keep the idea of a metric system but calibrate it to the fundamental constants of Nature. Would life not be easier if we called the Bohr radius a_0 just 1 atomic unit (au) of length rather than 5.2918×10^{-11} m even if that means that your height may be 34 GBohr (gigabohr)? Equally, if we get rid of our arbitrary daily life unit for energy, joule, then the number 4.3598×10^{-18} J (which is the natural combination $e^2/4\pi\varepsilon_0 a_0$ of fundamental constants) conveniently becomes 1 atomic unit of energy.

For some atomic units special names have been introduced, for example the atomic unit of length is *Bohr*, that of energy is *Hartree*. In this book we will use atomic units but we will mention kilojoules per mole (kJ/mol) where appropriate. Note that the atomic unit for charge is equal to the proton's charge e or 1.6022×10^{-19} C (so the charge of an electron in au is –1). It is popular to express the electron density in electrons per cubic Ångström (e/Å^3). To convert an electron density expressed in atomic units to e/Å^3 we multiply the value by 6.748.

Another advantage of atomic units is that equations become much simpler. For example, the Schrödinger equation in SI units for the hydrogen atom looks like this:

$$\left[-\frac{\hbar^2}{2m_e}\nabla^2 - \frac{e^2}{4\pi\varepsilon_0 r}\right]\psi = E\psi \tag{1}$$

whereas in atomic units it becomes:

$$\left[-\frac{1}{2}\nabla^2 - \frac{1}{r}\right]\psi = E\psi \tag{2}$$

are very fast. Since none of the results discussed in this book is obtained by these methods we will not discuss them here. The reader just needs to be aware that AIM uses the electron density ρ as the information source, which is defective if generated by semi-empirical calculations and which is *not* generated at all by molecular mechanics.

The practical boundaries of *ab initio* work are rapidly expanding via both the continual improvement of algorithms and spectacular hardware developments. For example, at the moment of writing the *ab initio* electron density of a protein with almost 700 atoms has been computed within an academic research environment. With the current hardware available we could easily triple the number of atoms if the military were interested in proteins. In summary, computational chemistry faces a bright future and its impact on the chemical community will undoubtedly increase. A full AIM analysis using a decent *ab initio* electron density has often been considered prohibitive by some but now presents itself as a rewarding and feasible option.

How can we explain briefly the chain of choices that have been made in solving the Schrödinger equation for the molecules and complexes that we discuss in this book? It is not our purpose to provide the reader with a full understanding of this chain from the information below. Rather we intend to show students (and novices) the wood for the trees in case they have already been exposed to the mathematical details of *ab initio* chemistry.

The first choice is the well known *Born–Oppenheimer approximation*, which treats the electrons as if they are moving in the field of fixed nuclei. This is a good approximation because, loosely speaking, electrons move much faster than nuclei and will almost instantly adjust themselves to a change in nuclear position. Anyone who has tried to rid their head of a swarm of mosquitoes will understand this approximation very well.

The second choice is the replacement of the *N*-electron wave function by a product of *N one*-electron wave functions or *spin orbitals*. A spin orbital is a wave function for an electron that describes both its spatial distribution and its spin (up or down). This product of one-electron wave functions is called a *Hartree product*. The price one pays for this simplification (which is due to Hartree) is that electron–electron repulsion is included only in an *average* way, in other words any electron experiences all remaining electrons via an average field. As a result the probability of finding one electron at a given point is independent of the position of another electron. According to probability theory, these two events are therefore *uncorrelated*. In reality, however, an electron instantaneously and directly repels another electron rather than via an average field. It is clear that the equations resulting from this approximation must be solved iteratively until a *self-consistent field* (SCF) is obtained. This is so because the average field appearing in each one-electron equation depends on the solutions of the other equations. This vicious circle can be cut by introducing an initial guess and iteration to a convergent solution.

We have, however, made one gross error when introducing the Hartree product because we did not take into account that electrons are actually indistinguishable. This means that electrons cannot be labelled. It is curious that one can label them anyway and correct for this mistake subsequently. Indeed, if electron 1 appears in spin orbital 1 and electron 2 in spin orbital 2, then we must equally allow for the event that electron 2 appears in spin orbital 1 and electron 1 in spin orbital 2. The correct wave function for these electrons is constructed from these two possibilities, because only then does it incorporate the fact that electrons are indistinguishable. It turns out that the correct wave function for electrons is *antisymmetric*. This means that swapping (or exchanging) the labels of any two electrons will merely cause the wave function to change sign. The wave function then obeys the *Pauli exclusion principle* and the (one-electron) Hartree equations become the *Hartree–Fock* (HF) equations. Correspondingly, the simplest antisymmetric wave function, which can be used to describe the ground state of an N electron system, is a single *Slater determinant*. The rows of such an $N \times N$ determinant are labelled by electrons and the columns by spin orbitals. Interchanging the coordinates of two electrons corresponds to interchanging two rows of the Slater determinant, which changes the sign of the determinant. A Slater determinant constitutes the third choice in the development of the most popular computational scheme in quantum chemistry today.

The Hartree–Fock equations are still complicated differential equations that even contain integrals. The fourth choice offers a nice way to tackle these integro-differential equations by expanding solutions of the HF-equations as linear combinations of basis functions, which form a set called the *basis set*. We then obtain matrix equations, which are called the *Roothaan–Hall equations*. They are solved *variationally*, which means that the linear combinations are 'varied' until the lowest possible energy is obtained, which then corresponds to the desired solution. Larger and larger basis sets will keep lowering the energy obtained via the *Hartree–Fock* approximation until a limit is reached, called the *Hartree–Fock limit*, E_0. As explained above, the motion of electrons is not correlated within the HF computational scheme, which is why E_0 is above the exact energy. The difference between these two energies is called the *correlation energy*. This energy is not a physical quantity which can be measured but a mathematical number resulting from the very nature of the HF approximation. Obtaining accurate estimates quickly of the correlation energy is one of the main challenges of computational chemistry in the 1990s.

Although many mathematical details have been omitted, we now have all the basic ingredients required to solve the Schrödinger equation for a single free atom (ignoring relativistic effects). For molecules we need one more key idea, which is due to Hund and Mulliken: *molecular orbitals* (MO), the fifth choice in our chain of decisions. Guided mainly by molecular spectroscopy, Mulliken introduced the drastic thought that orbitals are not just centred on one nucleus, but on every nucleus of a molecule. The resulting MOs are typically expressed as *linear combinations of atomic orbitals* (LCAO) which are in turn expanded in terms of (basis) functions in order to make practical calculations possible. Over the years, considerable effort has gone into the design of compact and reliable basis sets for each atomic number.

The sixth and final choice is the introduction of Gaussian basis functions, a bizarre proposal at first glance since they appear in the eigenfunctions of the Schrödinger equation for the harmonic oscillator rather than for the hydrogen atom. However, Gaussian basis functions make the computation of electron–electron repulsion integrals efficient, hence their position as a mainstream option in modern quantum chemistry. Many will have come across the special notations that have been developed to denote these Gaussian basis sets (for example, 6-31 + G**, STO-3G, LANL2DZ).

Putting all these choices together generates the impressive acronym *HF–SCF–MO–LCAO* to label the most typical route to the solution of the Schrödinger equation for molecules and aggregates of molecules. Clarifying in detail how $\rho(\mathbf{r})$ can be computed in terms of the wave function would require quite a few formulae with lengthy explanations as to what every symbol means. It is more important to understand the steps that have been taken to end up with $\rho(\mathbf{r})$ as a formidable expansion of Gaussian basis functions.

The Hartree–Fock scheme is remarkably successful in predicting the properties of systems near their ground-state equilibrium geometry, given that it ignores the correlation energy. Since the late 1980s, the quantum chemical community has become increasingly interested in an alternative to the then typical post-Hartree–Fock methods to tackle electron correlation. These post-Hartree–Fock methods include many-body perturbation theory (for example, Møller–Plesset), configuration interaction (for example, DCI, SDTCI), pair and coupled-pair theories (for example, CCSD). This alternative was found in a modern form of a theory called *density functional theory* (DFT).

These days it seems to be insufficiently recognized that the basic idea behind DFT is almost as old as quantum mechanics itself, dating back to work of Thomas, Fermi, Dirac and Wigner. This idea is that the energy of an electronic system can be expressed in terms of its electron density, i.e. without reference to orbitals. The original Thomas–Fermi model has been widely applied in atomic physics in spite of its highly approximate character. The subsequent Hartree–Fock–Slater or $X\alpha$ method originated in solid-state physics and evolved out of the need to develop techniques that were able to handle solids within a reasonable time.

In the mid-1960s, the work of Kohn, Hohenberg, Sham and others boosted the image and general usefulness of DFT. The well known *Hohenberg–Kohn theorems* basically guarantee that the exact energy of the ground state of a many-electron system can be expressed as a functional of ρ. At the time of writing this functional remains a holy grail, but the fact that it is known to exist brought about increasing research activity into better energy functionals. Kohn and Sham proposed introducing orbitals into the treatment of the kinetic-energy functional via the so-called *Kohn–Sham equations*. As a result, the kinetic energy could be computed simply and accurately and DFT became a practical tool for rigorous calculations. This important development is often misquoted as the birth of DFT, which is quite ironic because it introduces orbitals, which should be absent according to the original DFT spirit.

In the early 1990s, the quantum chemistry community embraced the now popular and economical DFT as a competitive alternative. In this book we also use it to include electron correlation into molecular wave functions. The community has formulated its own functionals, a survey of which is beyond the scope of this book. Unfortunately, improving energy functionals is not straightforward because the physical arguments which guided the development of early functionals appear to be exhausted. This is why, perhaps in surrender to this lack of guidance, recent research introduces further mixing of orbital-based considerations into functionals and even calibrates the functionals to experimental data. Nevertheless, all in all, DFT permits faster computations than the traditional methods. It also puts the spotlight on ρ rather than on the wave function, thereby inviting an interpretative electron density theory such as AIM.

All new examples (figures and tables) presented in this book are based on wave functions which include electron correlation. However, in the 1970s and 1980s, the theory of AIM was developed using Hartree–Fock wave functions, without electron correlation. We have deliberately avoided specifying all the details of the levels of calculation throughout this book for we do not want to deter the novice or bore the expert with jargon. However, we mention that optimizing molecules at Hartree–Fock level with the 6-31G* basis set is adequate, provided the corresponding electron density is replaced by one of higher quality (for the same nuclear skeleton). The level chosen for this electron density is the DFT method B3LYP (Becke–3–Lee–Yang–Parr) (Becke, 1993) and the Gaussian basis set is 6-311 + G(2d,p) (Foresman and Frisch, 1996) which encompasses both polarization and diffuse functions. The inclusion of electron correlation tends to move density away from the nucleus, thereby diminishing the values of the AIM atomic charges (see Chapter 9). This means that correlation tends to spread electronic charge more evenly over the atoms. It has also been observed that, for typical, stable molecules (first period elements), AIM properties show similar trends for second-order Møller–Plesset (MP2) and B3LYP electron densities versus Hartree–Fock densities. This means that a newly developed DFT method introduces correlation into ρ in a similar way to a traditional post-Hartree–Fock method.

The main point is that the theory of AIM starts from the electron density and does not care about how the density was obtained. At this point we should pause and think about the full impact of what it means to say that AIM concepts do *not* depend on the computational *scheme* chosen. This means that ρ is like a meeting ground, a platform of information generated by whatever computational scheme or method. This feature is one of the great advantages of AIM. Some concepts (such as population analyses) introduced by many popular models only exist within a given computational scheme, in other words they cease to exist if a molecular property is computed with a different method. Of course, the precise *values* of AIM quantities depend on the *level* of computation (for example, basis set) by which ρ is generated.

1.5 The electron density from experiment

One of the great advantages of using the electron density ρ as a source of information, as AIM does, is that an experimental electron density can be analysed just as well as a calculated one. The chemical insight gained by an

AIM analysis is equally valid for ρ retrieved from an X-ray diffraction experiment; although all examples we discuss in any later section of this book are based on computed electron densities.

Again, we will review a few important formulae and highlight a few key points. What we actually measure during a crystallographic experiment are structure factors $F(\mathbf{H})$ where $\mathbf{H}$ are indices denoting a particular scattering direction corresponding to a crystal plane. The electron density is obtained from these structure factors via the following equation, where V is the unit-cell volume:

$$\rho(\mathbf{r}) = \frac{1}{V}\sum_{\mathbf{H}} F(\mathbf{H})\exp(-2\pi i\mathbf{H}\cdot\mathbf{r}) \tag{1.1}$$

In the final stage of data processing, non-linear least-square refinements are used to fit a model electron density to the measured one. This is equivalent to minimizing the *difference density* $\Delta\rho$, which is defined as $\rho_{obs} - \rho_{calc}$ where the subscripts refer to 'observed' and 'calculated'. In routine molecular geometry determinations, ρ_{calc} is just a superposition of spherically symmetrical atomic densities. Most chemists are content with the determination of the nuclear skeleton of the molecule under study. In fact the nuclear positions are only measured indirectly via the electron density which is extremely high near the nuclei, except for hydrogen. Yet we know that the true molecular electron density is not just a superposition of chemically inert atoms.

Fortunately, an increasing number of affordable and high-quality diffraction apparatuses are now available, such as a low-temperature rotating anode or even a dedicated synchrotron facility with state-of-the-art area detector. These powerful devices yield an experimental electron density (ρ_{obs}) which contains much more information than required merely to find nuclear positions. In 1998 it was reported that a very high resolution picture of an amino acid could be obtained within one day (Koritsanszky *et al.*, 1998) whereas in the past this could be a matter of several months. In section 1.7 we outline how high-resolution crystallography typically extracts the extra information from ρ_{obs}, a procedure which is beyond bare molecular geometry determination. This book gives an alternative route.

1.6 Representing the electron density

It would be inconvenient to describe electron densities by means of endless tables of ρ evaluated at given points. If a picture is worth a thousand words it is definitely worth a million numbers. That is why we represent the electron density of methanal (formaldehyde) – the common thread throughout the book – via three different types of picture: a two-dimensional contour map, a three-dimensional contour map and a relief map. Each representation

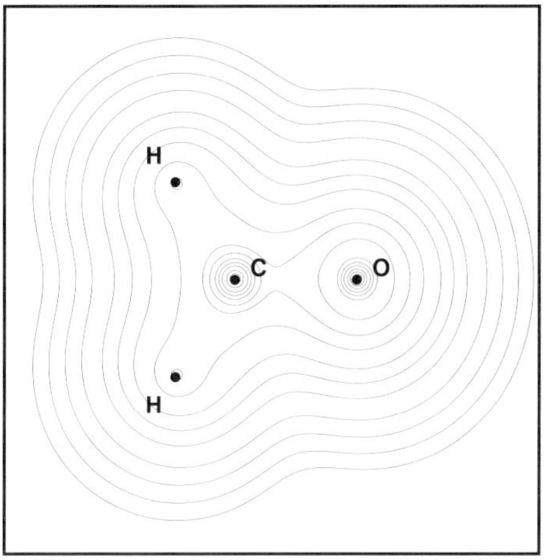

Figure 1.1 A contour plot of the electron density in the symmetry plane of methanal. The outer contour line corresponds to 0.001 au and the next contour lines correspond to values increasing according to the pattern 2×10^n, 4×10^n, 8×10^n where n varies from −3 to 2. This convention will be used throughout the book unless stated otherwise.

is in fact deficient because one needs four dimensions to display a 3D function completely. Hence, in the absence of colour it is rather difficult to produce a complete representation of $\rho(\mathbf{r})$.

Figure 1.1 shows the 2D contour map of ρ evaluated in the symmetry plane of methanal. The outer contour line (corresponding to $\rho = 0.001$ atomic units, au) can, for practical purposes, be taken as the outer limit of the molecule, although we will discuss this in more detail in Chapter 3. Inside this contour line lies a set of non-intersecting contour lines of higher electron density, each completely enclosing the next one (of higher ρ value). The latter property is abbreviated by stating that the contour lines are *nested*. The ρ values for each contour line increase according to *quasi*-geometrical sequence, which is given in the legend of Figure 1.1. We use these values throughout the book (unless stated otherwise). Note that at some point the curves become *disconnected*, i.e. they do not enclose the whole molecule any more. By drawing contour diagrams in other planes parallel to this one or in completely different planes, one can obtain a fair idea of the complete shape of $\rho(\mathbf{r})$.

Pictures such as in Figure 1.1 are in fact slices throughout a set of nested 3D *envelopes*, or contour surface. An envelope is a surface connecting points of the same electron density and an example is given in Figure 1.2 showing the $\rho = 0.001$ au envelope of methanal.

A different type of representation also yields 2D information but utilizes the third dimension to depict the value of ρ. We call this a *relief map* and provide an example in Figure 1.3. Just as in Figure 1.1, we show the electron

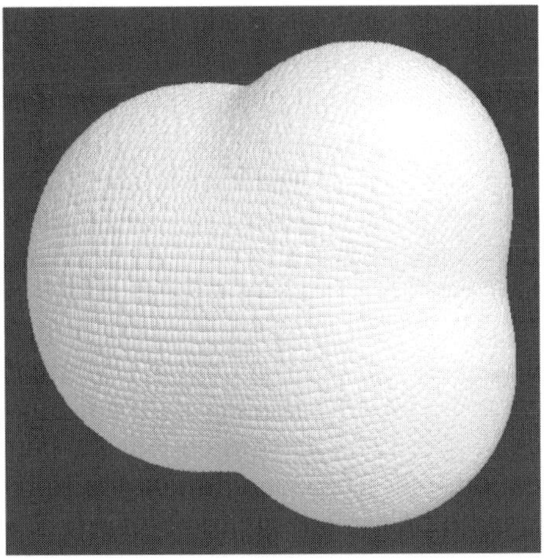

Figure 1.2 The 3D envelope of constant electron density (0.001 au) of methanal. This surface can be considered as the practical outer limit of the molecule. (*Technical remark*: For convenience this surface and more complicated topological objects throughout the book are represented by a multitude of little balls. Such surfaces should be thought of as being smooth.)

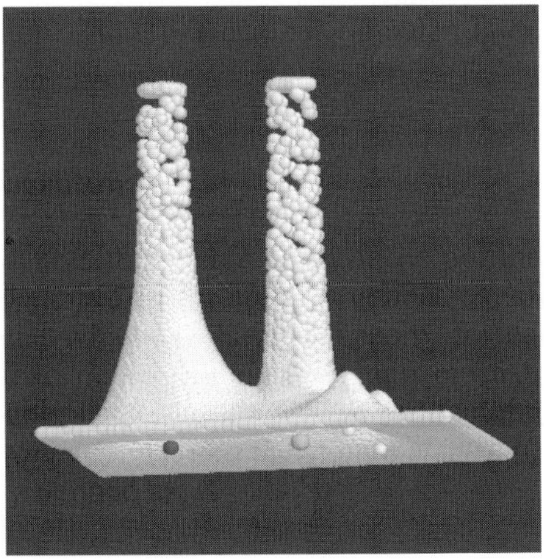

Figure 1.3 A relief map of the electron density in the symmetry plane of methanal. Electron density above 3 au is not shown. The density has been vertically accentuated by a factor of two for sake of clarity. The nuclei are marked under the relief map, oxygen (*left*), carbon (*middle*) and hydrogens (*right*).

density's profile in the symmetry plane of methanal. Its dramatic behaviour near the nuclei is quite obvious. At first sight it would seem that ρ is quite featureless except for indicating where the nuclei are: but is that really true? Is there any other information hidden in the electron density?

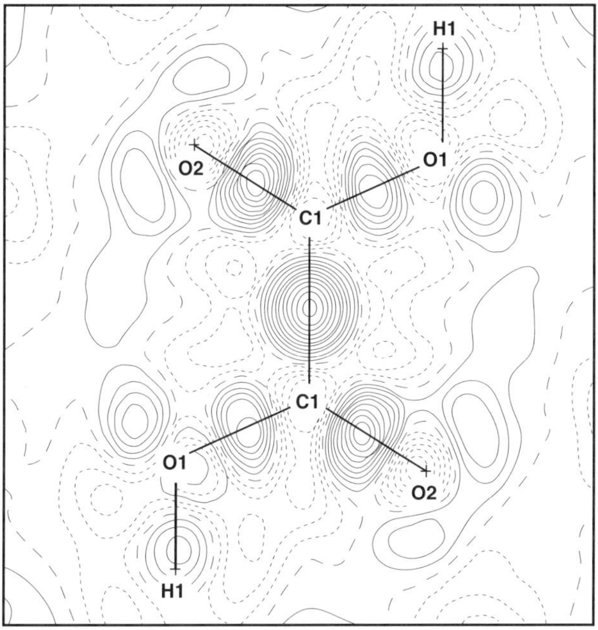

Figure 1.4 Standard deformation density in the molecular plane of the oxalic molecule at 15 K. Zero and negative lines are dashed. (From Zobel *et al.*, 1992. Reprinted by permission.)

1.7 Deformation densities

In the relief map shown in Figure 1.3 the electron density appears as a landscape of smooth background dominated by a handful of huge peaks. This is not just an artefact of the way a relief map represents the electron density. No matter how we look at ρ, the problem remains that somehow we have to eliminate the enormous contributions of the nuclear core regions in order to see the chemistry hidden in ρ. For that purpose, a few decades ago crystallographers introduced the *standard deformation density*, where the *promolecule density* is subtracted from the total density. The *promolecule* is a superposition of spherical ground-state atoms, centred at the nuclear positions. The promolecule density is a hypothetical reference density violating the Pauli exclusion principle because it is simply an ensemble of independent atoms prior to interatomic bonding, i.e. the density is not allowed to reorganize itself to comply with the exclusion principle. In spite of this defect it is still widely used in the construction of the deformation density. An example of a deformation density map for the oxalic acid molecule is given in Figure 1.4. It can be seen that the deformation density reveals features of chemical relevance such as bond regions and lone-pairs regions.

Regrettably, the price of invoking an arbitrary, hypothetical reference density is paid at some stage. For example, there is a lack of density in the O–O bond in hydrogen peroxide and similar problems arise in the fluorine molecule. In general, electron density accumulation in bonds in a deformation map is indicative of covalent bonding, but the opposite is not true, i.e. some covalent bonds do not show an accumulation of electron density. Attempts have been made to resolve these inadequacies by proposing modified reference densities which are perhaps even more artificial. Rather than proposing alternative electronic configurations of 'promoted' atoms reflecting their 'preparedness' to bind, we question the validity of invoking a reference density in the first place. If we can extract chemistry from the electron density itself then why not apply Occam's razor and discard the idea of an external reference density altogether?

This is exactly what some crystallographers are starting to do using AIM. A *topological analysis* of the electron density or AIM analysis puts great demands on the quality of an experimental electron density, but such densities can now be routinely obtained.

Summary

- Quantum mechanics provides us with a consistent and powerful mathematical formalism (rules and equations) to predict any property of a molecule. Nevertheless, a complete understanding and clear interpretation of these rules and equations is still unavailable.
- The Schrödinger equation for large molecules can now be routinely solved *ab initio* by means of computer programs. This is done via a computational scheme that can be summarized by the acronym HF–SCF–MO–LCAO supported with Gaussian basis sets and density functionals.
- The electron density ρ can be measured experimentally and computed by a large variety of computational schemes. It is therefore an ideal 'meeting ground' or 'communication platform' on which to base an interpretational theory such as AIM.
- The theory of 'atoms in molecules' takes advantage of the electron density as an information source from which to (re)formulate chemical concepts. It thus provides a bridge between chemistry and quantum mechanics.
- Deformation densities have been designed to eliminate the dominance of the nuclear core contributions to the density and so to reveal chemical features, but they have their limitations. AIM suggests that there is no need to invoke arbitrary reference densities: all chemistry is already hidden in ρ.

References

Becke A. D. (1993). Density-functional thermochemistry. III: The role of exact exchange. *Journal of Chemical Physics*, **98**, 5648–5652.

Cramer J. G. (1986). The transactional interpretation of quantum mechanics. *Reviews of Modern Physics*, **58**, 647–687.

Foresman J. B. and Frisch A. (1996). *Exploring Chemistry with Electronic Structure Methods* 2nd edn. Pittsburgh: Gaussian Inc.

Koritsanszky T., Flaig, R., Zobel D., Krane H-G., Morgenroth, W. and Luger P. (1998). Accurate experimental electronic properties of DL-proline monohydrate obtained within 1 day. *Science*, **279**, 356–358.

Zobel D., Luger P., Dreissig W. and Koritsansky T. (1992). Charge density studies on small organic molecules around 20 K: oxalic acid dihydrate at 15 K and acetamide at 23 K, *Acta Crystallographica*, **B48**, 837.

Further reading

Bethe H. (1964). *Intermediate Quantum Mechanics*, Lecture Notes and Supplements in Physics. New York: W. A. Benjamin.

Coppens P. (1997). X-ray charge densities and chemical bonding. *IUCr Texts on Crystallography* Vol. 4. Oxford: Oxford Science Publications.

Feynman R. P., Leighton R. B. and Sands M. (1963). *The Feynman Lectures on Physics*. Wokingham: Addison-Wesley.

Hehre W. J., Radom L., Schleyer P. v. R. and Pople J. A. (1986). *Ab Initio Molecular Orbital Theory*. Chichester: Wiley.

Hinchliffe A. (1996). *Modelling Molecular Structures*. Chichester: Wiley.

Holland P. R. (1993). *The Quantum Theory of Motion: An Account of the Broglie–Bohm Causal Interpretation of Quantum Mechanics*. Cambridge: Cambridge University Press.

Kryachko E. S. and Ludeña E. V. (1990). *Energy Density Functional Theory of Many-electron Systems*. Dordrecht: Kluwer Academic.

McWeeny R. (1992). *Methods of Molecular Quantum Mechanics* 2nd edn. London: Academic Press.

Moore W. (1993). *Schrödinger: Life and Thought*. Cambridge: Cambridge University Press.

Nebeker F. (1995). *Calculating the Weather: Meteorology in the 20th Century*. London: Academic Press.

Parr R. G. and Yang W. (1989). *Density-functional Theory of Atoms and Molecules*. Oxford: Oxford Science.

Smith S. J. and Sutcliffe B. (1997). The development of computational chemistry in the United Kingdom. In *Reviews in Computational Chemistry* (Lipkowitz K. B. and Boyd D. B., eds), Vol. 10. Cambridge: VCH Publishers.

Szabo A. and Ostlund N. S. (1982). *Modern Quantum Chemistry: Introduction to Advanced Electronic Structure Theory* rev. edn. New York: McGraw-Hill.

Ziegler T. (1991). Approximate density functional theory as a practical tool in molecular energetics and dynamics. *Chem. Reviews*, **91**, 651–667.

Chapter 2

The gradient vector field

Dad, why are you not holding my hand any more?
You'll be fine, son, just follow your own sense

2.1 What is the gradient?

It is clear from the previous chapter that the molecular electron density is dominated by huge values near the nuclei. In order to eliminate these nuclear peaks, crystallographers have introduced the deformation density. In many cases this density reveals chemical features such as bonding regions and lone pairs. Still the main question remains: can we disclose chemistry in the molecular electron density without subtracting from it a reference electron density? The answer is yes. In fact we do not need to introduce an external reference density at all: we can use the original molecular electron density itself. In other words, we eliminate the need for an extra arbitrary reference density to compare with our original molecular electron density. A mathematical tool called the *gradient vector* or simply the *gradient* makes it possible to invoke the molecular density as its own reference.

The gradient is defined via partial derivatives usually in the context of a mathematical branch called vector calculus (see Box 2.1). The following rather exotic but simplified illustration of the gradient will help to understand the physical meaning behind its mathematical definition.

Suppose a scientist is aboard a small submarine somewhere in the middle of the Atlantic Ocean. She is exploring the bottom of the sea near the mid-Atlantic fault where ocean water is expected to be in contact with slowly emerging lava. Suppose that the submarine's spotlights do not reach further than a few metres in the pitch-dark water but that it is equipped with thermometers all around. At a distance of several tens of metres the scientist is still able to detect the place where lava emerges from the ocean floor simply by reading the water temperature in various directions around the submarine. This is possible because the temperature decreases away from this hot spot.

Put more formally, we know that the fault is surrounded by a nested set of envelopes or surfaces of constant temperature. As the scientist is interested in finding where the lava emerges, she will travel towards the higher

Box 2.1 A closer look

The gradient

The gradient (vector) of a 3D scalar function $\varphi(x, y, z)$ is defined as follows:

$$\nabla\varphi \equiv \mathbf{u}_x \frac{\partial\varphi}{\partial x} + \mathbf{u}_y \frac{\partial\varphi}{\partial y} + \mathbf{u}_z \frac{\partial\varphi}{\partial z}$$

where $\mathbf{u}_x$, $\mathbf{u}_y$ and $\mathbf{u}_z$ are the three unit vectors. It can be proven that if $\varphi(x, y, z) = c$ is the equation of a surface, then $\nabla\varphi$ evaluated at some point in space is normal to this surface at that point. This statement makes sense only if $\nabla\varphi$ is different from the null vector. We have used this property to introduce the gradient in the text. Another important property is that the gradient always points in the direction of greatest increase in φ. Finally, the sense of the gradient vector is towards the contour surface with the *higher* function value.

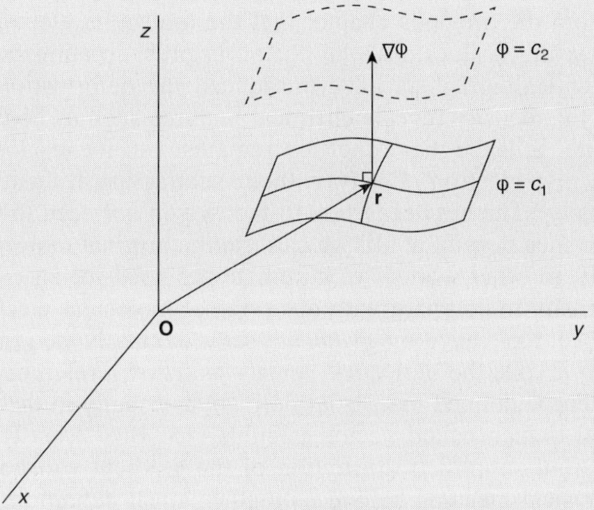

Figure 1 A simplified diagram illustrating the gradient vector $\nabla\varphi$ evaluated in a point described by the position vector **r**. The gradient is perpendicular to the surface of constant φ value (c_1) and it is orientated to a surface with a higher φ value (c_2).

temperature regions. However, the quickest way to find the fault is to move in the direction in which the temperature rises most. This direction corresponds exactly to the gradient vector because it is this vector that points in the direction of greatest increase in the scalar (function) at hand, which is the temperature in our case. Another property of the gradient vector is that it is everywhere perpendicular to an envelope of constant scalar value (i.e. temperature in this example).

Returning to the electron density we can similarly say that the gradient vector of the electron density, denoted by $\nabla\rho$, points in the direction of greatest increase or steepest ascent in ρ. Also, because ρ is just another 3D scalar function like temperature, $\nabla\rho$ is everywhere perpendicular to a surface of constant electron density.

There are many other physical or daily life situations that illustrate the gradient vector. Perhaps the most familiar example is that of the direction of steepest ascent on a hill. The shortest path to the top can be found by tracing a curve everywhere orthogonal to the pattern of height contour lines on a map. Another example is the wind: the gradient of the air pressure coincides with the local wind direction (ignoring the Coriolis effects). As another example we consider an object originally at rest falling onto the earth's surface. In this case the scalar function is the earth's gravitational potential and the corresponding gradient vector is tangential to the object's path.

In summary, there is nothing puzzling about $\nabla\rho$ because there are many familiar or easily imaginable physical situations that share the properties of $\nabla\rho$. We just have to replace ρ by another scalar function closer to human experience in order to understand $\nabla\rho$ better. Box 2.1 furnishes a formal definition and review of the gradient vector.

2.2 The gradient path

As in the previous section, we introduce the concept of a gradient path first in terms of its physical meaning before formulating the mathematical definition. The example of the falling object, initially at rest, is a good case to illustrate what a gradient path is. We asserted before that at any time the gradient vector is tangential to the path followed by the object. Here the direction of the gradient vector actually coincides with the direction of the force acting on the object. Of course, the two vectors – the gradient and the force – are oriented differently because the force **F** is equal to minus the gradient of the gravitational potential V, or $\mathbf{F} = -\nabla V$. Because of this correspondence the path followed is known as the *line of force*. An equivalent name for such a gradient path is *field line*.

Another classical example illustrating the meaning of a gradient path is that of field lines appearing in electrostatics. Figure 2.1 shows two identical charges separated by a given distance and generating an electrostatic potential which is represented by contour lines. These contour lines are actually the intersections of the isopotential envelopes and the plane of the page. It is clear that there are curves terminating at either charge which are everywhere orthogonal to the contour surfaces. These curves are known as field lines and are in fact the gradient paths associated with the potential. They have a clear interpretation because a positive test charge released somewhere in the system would move along these field lines. The force acting on such a particle is tangential to the gradient path and points in the direction opposite

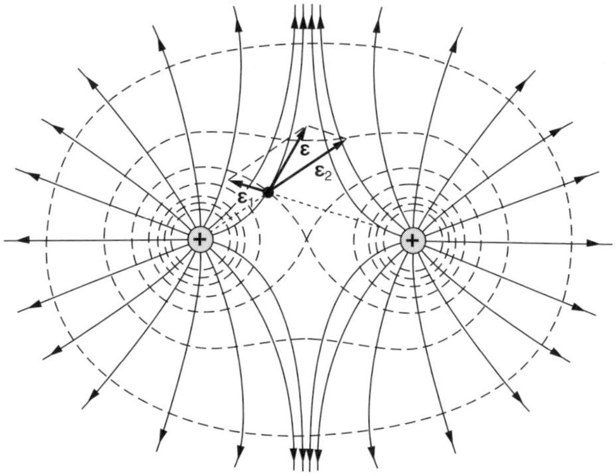

Figure 2.1 Lines of force and equipotential surfaces (dashed curves) of the electric field of two identical charges. (From Alonso and Finn, 1983. Copyright © Addison-Wesley Publishing Company; reprinted by permission.)

to the gradient vector because, again, $\mathbf{F} = -\nabla V$ where V is the electrostatic potential. Because of this minus sign the positive test charge will move away from the terminus of the gradient path it is advancing on. It is important to realize, however, that gradient paths corresponding to $\nabla\rho$ must *not* be interpreted in terms of an actual force. The analogy between an electron density gradient path and these examples is purely formal.

The precise definition of a gradient path is given in Box 2.2, but for a less rigorous understanding it suffices to think of the following construction. Evaluate $\nabla\rho$ in a given point and follow this vector over a very small stretch. At the end-point of this vector, re-evaluate $\nabla\rho$ and again follow the vector over a very short segment and so on. This sequence of vector segments forms a piecewise continuous trajectory, which approaches the gradient path in the limit of infinitesimally small vector segments. This procedure is made clear in Figure 2.2.

Three important properties of gradient paths should be mentioned. The most obvious property of a gradient path is that they are everywhere orthogonal to surfaces of constant value in the scalar function. Since a gradient path is basically a succession of tiny segments of gradient vectors, it adopts this property from the gradient vector itself. This property is illustrated in Figure 2.3.

The second property is that there is only one gradient path going through a given point provided that at that point $\nabla\rho \neq \mathbf{0}$. This is equivalent to stating that gradient paths never cross except when $\nabla\rho = \mathbf{0}$. This property results from the fact that there is only one gradient at any given point, i.e. $\nabla\rho$ is uniquely defined at a given point. Indeed, if gradient paths could cross then

Box 2.2 A closer look

The gradient path

A *gradient path* is defined as a curve to which the gradient vector is tangential at each of its points (where the gradient does not vanish). Practically speaking, take a gradient path, pick a point on this path and draw the tangent to it through this point. You will then recover the gradient vector on this tangent (through the given point). Figure 1 helps to cast this statement into a mathematical formula. If ℓ represents the path length, $\mathbf{r}$ is the position vector of an arbitrary point on the gradient path and $|\nabla\rho|$ is the magnitude of the gradient vector, then the following equation defines a gradient path:

$$\frac{d\mathbf{r}}{d\ell} = \lim_{\Delta l \to 0} \frac{\mathbf{r}(\ell + \Delta\ell) - \mathbf{r}(\ell)}{\Delta\ell} = \frac{\nabla\rho}{|\nabla\rho|}$$

This is in fact a system of three ordinary differential equations which is very hard to solve analytically because of the complicated nature of ρ. Instead we resort to numerical integrators such as the Runge–Kutta–Cash–Karp method. The most elementary way to solve these equations is the Euler method which is described in the text (see also Figure 2.2). It is also possible to find an analytical expression for a gradient path but the details are beyond the scope of this book.

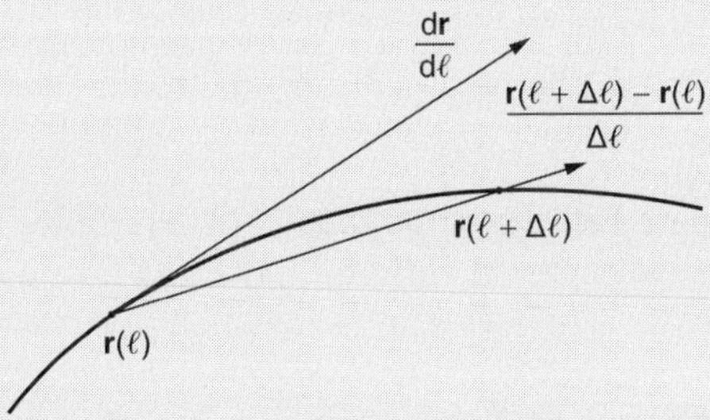

Figure 1 A curved gradient path is plotted (bold line) as a set of position vectors $\mathbf{r}$ depending on a path parameter ℓ which coincides with the path length. The vector intersecting the gradient path in two points with position vector $\mathbf{r}(\ell)$ and $\mathbf{r}(\ell + \Delta\ell)$ becomes the unit tangent vector $d\mathbf{r}/d\ell$ in the limit when the two points coalesce. The normalized gradient vector $\nabla\rho/|\nabla\rho|$ coincides with the unit tangent vector to the path at point $\mathbf{r}$. Since this is true for any point along the path (provided $\nabla\rho \neq \mathbf{0}$), the path can be defined by this property.

there would be two different gradient vectors at the intersection point (each one tangential to a different path), which is impossible.

The third property is that gradient paths have a beginning and an end. This is because a vector has an orientation (which can be from left to right or reverse) and a gradient path is built up by tiny segments of gradient vectors

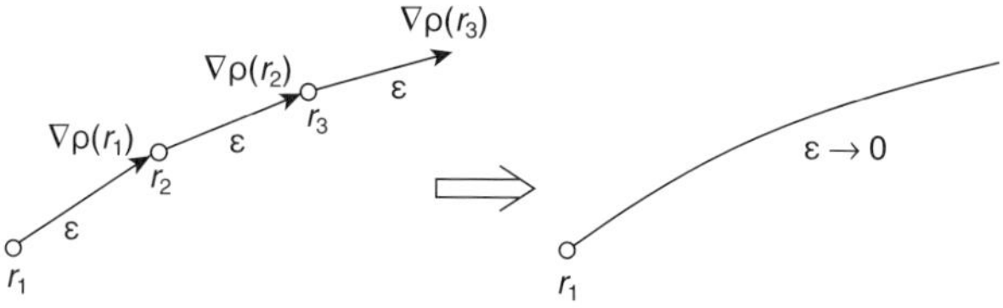

Figure 2.2 The construction of a gradient path as a succession of infinitesimally small gradient vector segments (*left*). The left illustration is a piecewise continuous approximation of the actual gradient path (*right*).

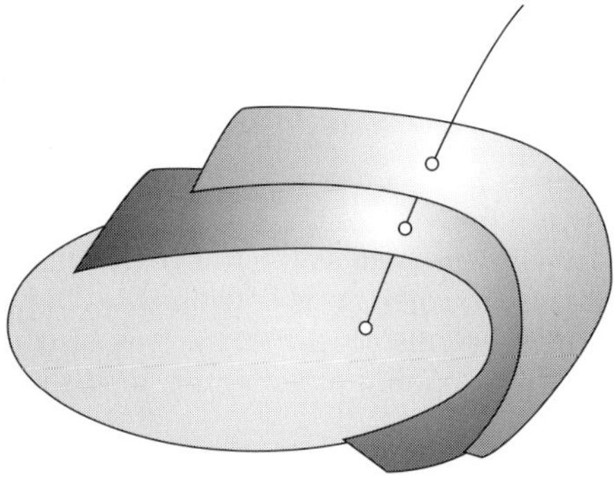

Figure 2.3 A gradient path traversing a set of nested envelopes or surfaces of constant electron density. The path is orthogonal to all envelopes.

which are all orientated in the same way. For example, in a free molecule most gradient paths originate at infinity and most terminate at a nucleus.

2.3 The gradient vector field

The *gradient vector field* is in essence an infinite collection of gradient paths. As an example we show the gradient vector field of ρ in the molecular plane of methanal in Figure 2.4. This figure only shows a finite number of gradient paths, enough to render a sufficiently detailed representation of the actual gradient vector field. Recalling that only one gradient path passes through a given initial point, how do we select a set of initial points that yield a good representation of a complete vector field? In practice, we trace the gradient paths *backwards* in the direction of decreasing ρ. We start from an equally spaced set of initial points on a small circle about any of the four nuclei and trace the paths away from the respective nuclei.

The full gradient vector field (outside the molecular plane) would be obtained by selecting a large number of initial points on a small sphere about every nucleus. In that sense the collection of gradient paths in the molecular

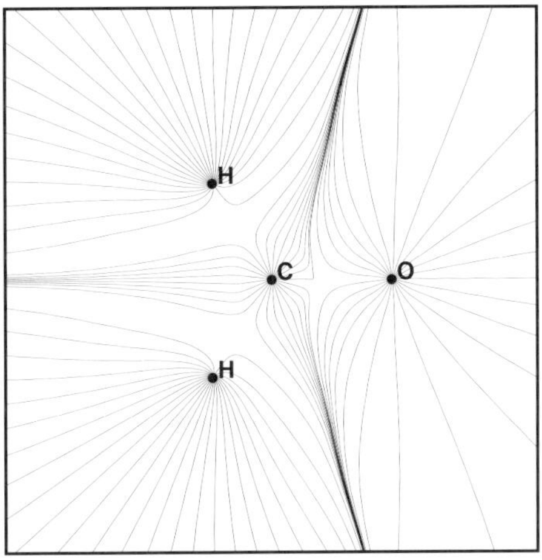

Figure 2.4 A map of the gradient vector field of the electron density for the plane containing the nuclei of the methanal molecule. Each line represents a gradient path of $\nabla\rho$. Only the paths which terminate at the positions of the nuclei are shown. Each path is arbitrarily terminated at the surface of a small circle about a nucleus. In practice, however, the paths are traced starting from this circle away from the nucleus.

plane is a subset of the total gradient vector field. We could regard the former as the intersection of the total gradient vector field with the molecular plane.

Figure 2.5 shows the planar subset of the gradient vector field again but now superimposed on a contour map of ρ. It is clear that the gradient paths are everywhere orthogonal to the contour lines, a property which is valid in three dimensions as well. The contour lines on one hand and the gradient paths on the other can be looked upon as two mutually orthogonal families of curves. They both describe the electron distribution in a complementary way.

It is clear from this example that gradient paths do not intersect except where they meet such as at the nuclei. As explained above, we correctly infer from this observation that at the nuclei the gradient must vanish. Indeed, the electron densities which are routinely computed, such as the ones we discuss throughout this book, all have zero gradient at the nuclear positions. However, it has been proven that the exact (rather than our usual approximate) electron density shows a cusp at the nucleus. Since at a cusp the gradient (or derivative) is not defined, we cannot, strictly speaking, corroborate the inference made above.

The small arrows in Figure 2.5 remind us that the gradient paths have a sense. All paths shown terminate at a nucleus and are thus said to be *attracted* to this nucleus which is therefore called a *nuclear attractor* (NA). Every nucleus dominates a portion of space by attracting gradient paths towards itself. For example, in Figure 2.5 the oxygen nucleus controls the

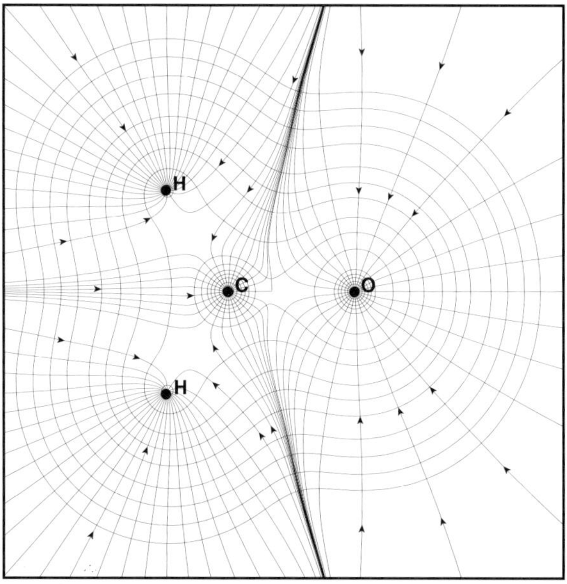

Figure 2.5 The map of the gradient vector field in Figure 2.4 superimposed on a ρ contour map for the molecular plane of methanal. Notice that the gradient paths and the contour lines constitute a mutually orthogonal family of curves. Some gradient paths have been given arrows to mark their sense.

right-hand region, which consists of points through which gradient paths pass attracted to the oxygen nucleus. In this sense the molecule seems to fall apart into disjointed regions each containing one nucleus. In other words, the gradient vector field naturally partitions the molecular electron density into regions which we will later recognize as the atoms that make up the molecule.

Consider two neighbouring nuclei in methanal such as carbon and oxygen shown in Figure 2.6. Depending on the choice of initial point we can make a gradient path terminate at the carbon nucleus or at the oxygen nucleus. This procedure triggers the question, can we find a gradient path that does not terminate at either nucleus but somewhere in between? After all, it is intuitively clear that there must be a boundary between the two atoms which is most likely to consist of gradient paths itself.

2.4 Critical points

Let us look at Figure 2.6 again. We conjecture that there is a gradient path (path A) heading for the molecule from infinity *without* terminating at the carbon nucleus or the oxygen nucleus. Assuming a path of that property exists, we ask ourselves in which point it will terminate. Furthermore, if this

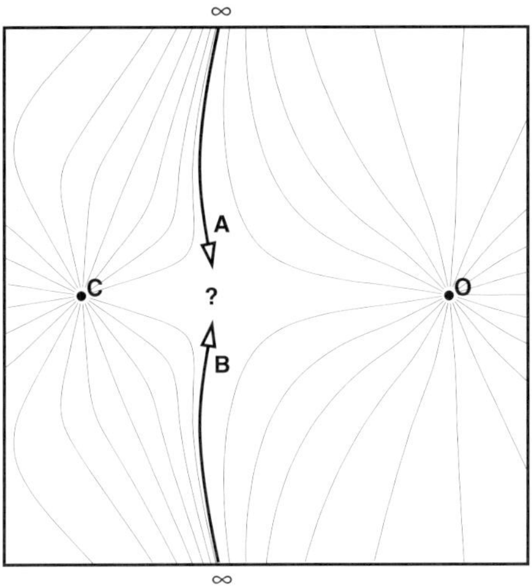

Figure 2.6 An enlargement of the gradient vector field between the oxygen and carbon nucleus in methanal. The bold gradient paths A and B, which originate at infinity, are not attracted to either nucleus. They will meet each other somewhere between the two nuclei. At the point where they meet the gradient vector must vanish. A point where the gradient vanishes is called a critical point.

path exists there must an alternative gradient path (called B) sharing this property simply because of symmetry. These two paths A and B will then intersect at some point which is impossible, unless the gradient vanishes at this intersection point or $\nabla\rho = \mathbf{0}$. It turns out that there is indeed such a remarkable point between the oxygen and carbon nuclei. It is called a *critical point* (CP) because the gradient vector is zero at this point.

We have already encountered another type of CP, namely the nuclear attractor. Recall that a host of gradient paths meet at a nucleus which is possible only if $\nabla\rho = \mathbf{0}$. So the nucleus is also a CP or, to be more precise, there is a CP in ρ which practically coincides with the nucleus. There are two further types of CP, as we will demonstrate in Chapter 5.

It is helpful to introduce another analogy to illustrate the concept of a CP. In fluid mechanics one frequently uses the terms *velocity potential* and *stream function*. The former can be identified with the scalar function φ, for example, the temperature or the electron density. The latter can be identified with the gradient vector field corresponding to φ (for example, lines of force or our $\nabla\rho$ field). This concept is easier to interpret physically because the velocity of a fluid element is simply $-\nabla\varphi$.

Now consider rafting on a mountain river. Presuming that the river contains incompressible, non-viscous and circulation-free water (the latter is not a very good approximation) the quantity $-\nabla\varphi$ can be associated with the velocity vector of the raft. Calm waters match regions where $\nabla\varphi$ is small and perfect backwater would map onto CPs in φ. Realizing that high values of $|\nabla\varphi|$ correspond to the rapid flow in a cascade, one can extend the analogy with $\nabla\rho$ even further. Travelling down a cascade would then correspond to moving in the vicinity of a nucleus because there the gradient of the electron density is very large (before it eventually turns to zero at the nucleus).

There is one more interesting comment on a feature of gradient paths in connection with CPs. For a free or isolated molecule, the magnitude of $\nabla\rho$ decreases to very small values in the outer regions surrounding it. In the limit we find that $|\nabla\rho| = 0$ in any point infinitely far removed from the molecule. This is equivalent to declaring that there is an infinite set of CPs at infinity. Since this is where many gradient paths originate, and knowing that all gradient paths terminate at a CP, we conclude that a gradient path always links two CPs.

2.5 Some examples of gradient vector fields

We will now look at the gradient vector field of a set of four molecules drawn from various chemical disciplines. This set encompasses the carbonate ion CO_3^{2-}, the organometallic catalyst $CH_3TiCl_2^+$, the amino acid glutamine, the complex $(H_2O)_2$ and a piece of the molecular crystal of chlorine. The variety of this set illustrates that the current analysis knows no boundaries across chemistry. The potential applications range from biochemistry to organic, inorganic, physical and solid-state chemistry. By no means can we represent these vast subdisciplines by a mere handful of systems such as the present set. What is demonstrated, however, is that once a reasonably accurate experimental or theoretical electron density is known for any system, the gradient vector field can be obtained from which, as we shall see, a wealth of chemical information can be derived.

Figure 2.7 shows a superposition of the gradient vector field and the contour map for the carbonate ion CO_3^{2-} in the plane of all the nuclei. A comparison with an analogous picture for methanal (Figure 2.5) reveals interesting similarities. The portion of space that the carbon nucleus dominates in methanal has more-or-less the same shape as the corresponding portion of space in the carbonate ion. This region of space reflects the trigonal carbon appearing in both molecules. The two regions dominated by the oxygen nuclei that replace the hydrogen nuclei in methanal take more space than their hydrogen counterparts. As a consequence, the outer regions of the carbon are pushed towards the carbonyl oxygen.

Finally, an important phenomenon that appears in both molecules (and in fact in any molecule) is that the gradient paths are most curved between the nuclei and are almost straight lines in the vicinity of the nuclei. It is tempting to explain this observation as follows: the gradient vector field directly expresses the regions inside the molecule where chemistry is at work, i.e. *between* two bonded nuclei not *at* the nuclei. After all, when free atoms join

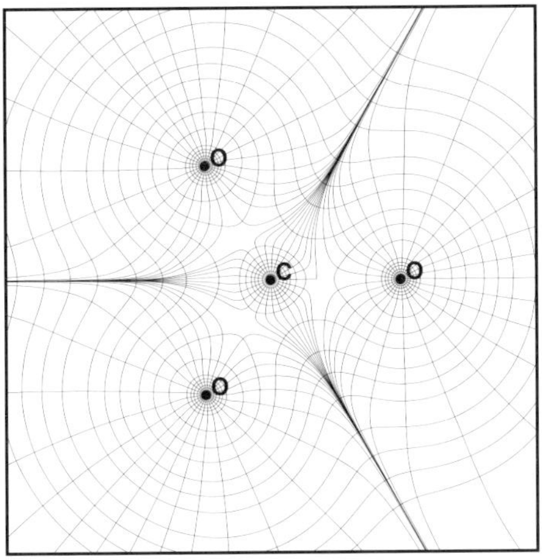

Figure 2.7 The map of the gradient vector field superimposed on a contour map of the electron density for the molecular plane of the carbonate ion. It is instructive to compare this map with the one in Figure 2.5 as the hydrogens are replaced by oxygens.

to form a molecule it is the outer regions which are affected most, hence the electron density in the immediate neighbourhood of the nucleus hardly changes regardless of its chemical environment. Therefore the space in the vicinity of the nucleus does not reveal any details of how one nucleus is bonded to another. This is precisely reflected in the gradient vector field, which shows the same radial pattern in the proximity of any nucleus independent of its chemical environment. On the other hand, the pattern of the gradient vector field in the region *between* nuclei is very different from that of the vector field of free atoms, because in this region the *interaction* between atoms is expressed.

All the Figures (2.7–2.10) are plotted in a 10×10 au window to allow easy inspection and comparison amongst each other.

The second example, $CH_3TiCl_2^+$, is displayed in Figure 2.8 where we plot the gradient paths in the mirror plane of the molecule that contains the titanium, carbon and one hydrogen nucleus. Again, because of symmetry all the gradient paths stay in the plane provided their initial points are also in the plane. Note that the region dominated by the titanium nucleus is large.

The third example (glutamate) is displayed in Figure 2.9. For the first time we show a molecule without symmetry. In view of the absence of symmetry in this conformation of glutamate, there is no collection of gradient paths confined to within a symmetry plane. As a result, many gradient paths shown in the plotting plane of Figure 2.9 seem to end prematurely. In reality the paths which are traced backwards from circles of initial points lying in the plotting plane *leave* the plane at some point and continue on in the third dimension. This artefact is actually more dramatic because we have

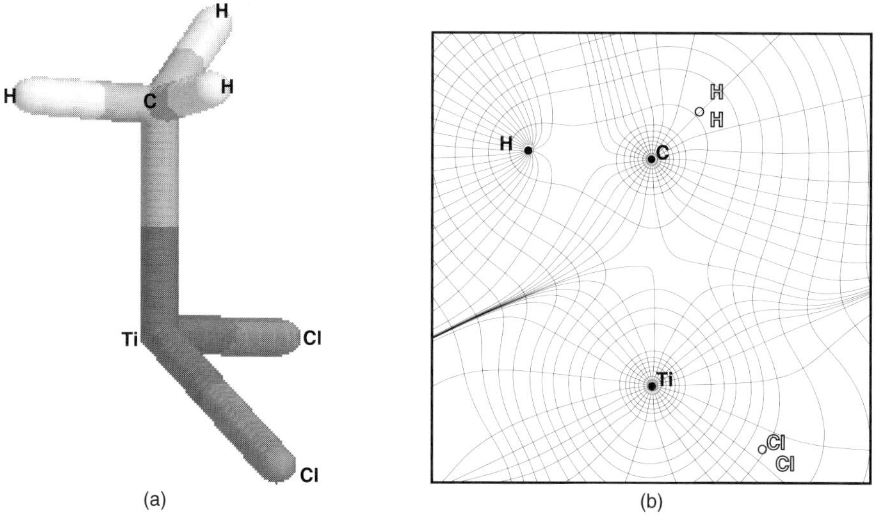

Figure 2.8 (a) Stick diagram of a conformation of $CH_3TiCl_2^+$. (b) Map of the gradient vector field superimposed on a ρ contour map in the mirror plane of $CH_3TiCl_2^+$. The chlorine nuclei are marked by an open letter type to indicate that they lie outside the plotting plane.

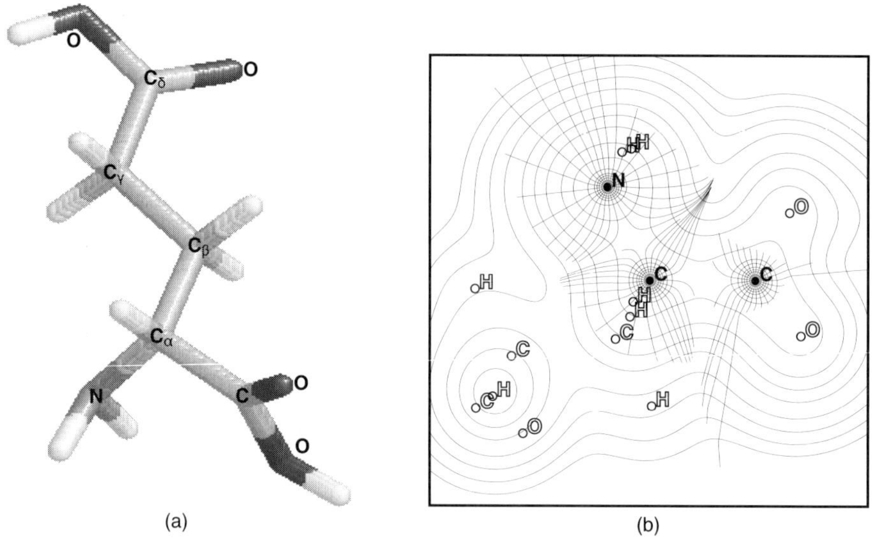

Figure 2.9 (a) Stick diagram of a conformation of the amino acid glutamate. (b) Map of the gradient vector field superimposed on a contour map of the electron density in the C–C_α–N plane of glutamate. The atoms with open labels do not lie in this plane.

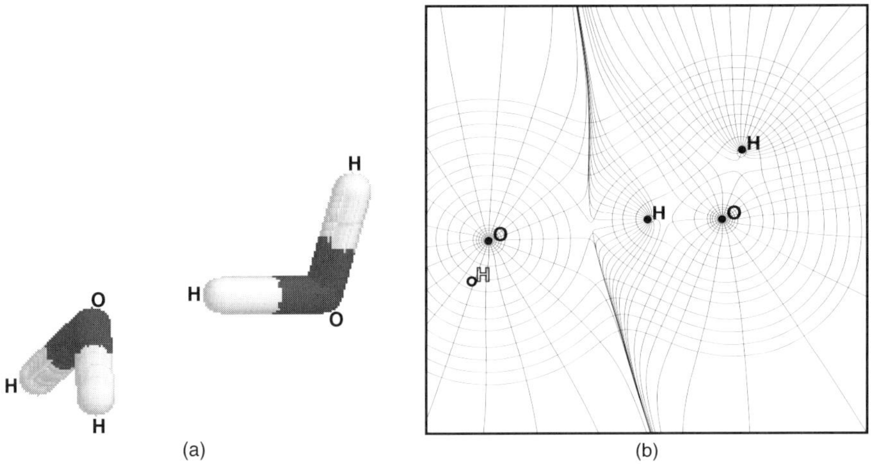

Figure 2.10 (a) Stick diagram of a conformation of the water dimer. (b) Map of the gradient vector field superimposed on a contour map of the electron density in the water dimer. The plotting plane encompasses the water molecule on the right and the oxygen nucleus of the left water molecule.

attenuated it here by projecting the paths into the plane unless they are separated from it by more than 0.2 au.

The fourth example is the water dimer $(H_2O)_2$ shown in Figure 2.10. The main reason for including this example is to establish that gradient vector fields also apply to complexes or aggregates of molecules. It is intuitively clear from this picture that there is a sharp boundary between the two water molecules. Again the system has partitioned itself in a natural way. We have not imposed any boundaries on the system; we have not drawn any arbitrary lines or planes to slice the dimer into two molecules or to cut the molecule further into separate atoms. This idea can be extended by considering not just a dimer but an immense aggregate of molecules such as a solid.

Solid chlorine is a molecular crystal which consists of Cl_2 molecules that form layers parallel to the so-called (100) plane (or *bc* plane). Figure 2.11 represents such a layer which coincides with a mirror plane perpendicular to the *a*-axis. This figure contains some details that are not relevant at the moment. The main point is that once a sufficiently reliable electron density of any system has been obtained, the gradient vector field provides the basic information we need to characterize the chemistry of that system. How this is done will become clear in forthcoming chapters. Note that the regions traversed by gradient paths attracted to any chlorine nucleus do not extend to infinity as in the case of free molecules.

The gradient vector field may seem a rather abstract concept remote from any real chemistry. We hope that the following chapters will show this impression to be quite wrong because the gradient vector field is the

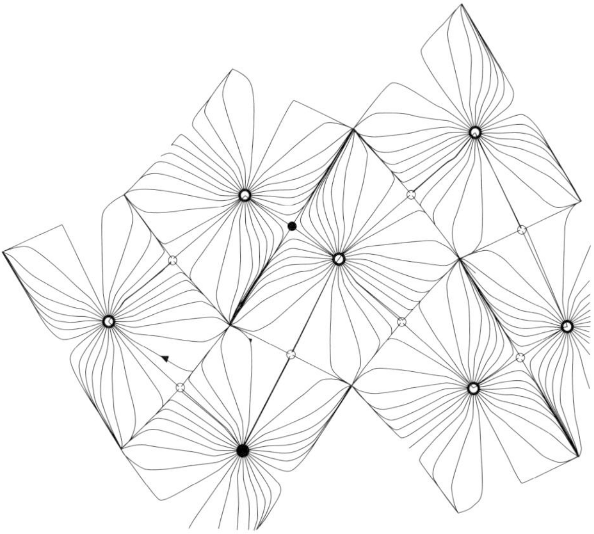

Figure 2.11 The gradient vector field in the (100) plane of solid chlorine. (From Tsirelson, *et al.*, 1995.)

foundation on which the theory of atoms in molecules rests. Explaining this concept carefully is entirely justified in the light of what follows. Indeed, two cornerstones of chemistry, the atom and the bond, will be defined using the gradient vector field only.

2.6 Non-nuclear attractors

It is possible that gradient paths are attracted to a point in space where one would expect (by inspection of the gradient vector field) a nucleus to reside, only to perceive that it is absent. An example is given in Figure 2.12. The molecule Li_2 is rather unstable and is detected only in 1% concentration in lithium vapour at its boiling point. Nevertheless a wave function for it can be readily computed and analysed in terms of the gradient of the electron density. We conclude that there must be a CP exactly between the Li atoms because gradient paths are meeting there. Such a CP is called a *non-nuclear attractor* (NNA). We will see later that this CP is a maximum of ρ in 3D, a property it shares with the nuclear attractor. So, topologically the nuclear and the non-nuclear attractor are identical which is why we propose the acronym (N)NA to introduce a proper name for the CP which is a maximum in ρ. The (N)NA notation expresses compactly the merging of the acronyms NA and NNA, which cannot be distinguished topologically.

The revealed topological pattern persists in very accurate wave functions in lithium clusters up to Li_6 (Gatti *et al.*, 1987). It is now believed that non-

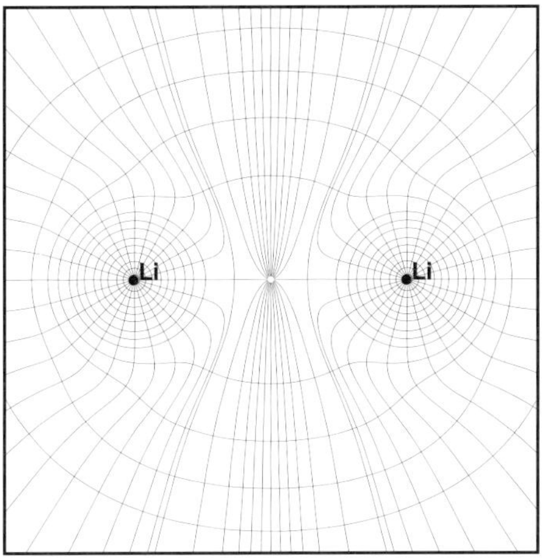

Figure 2.12 The map of the gradient vector field superimposed on a contour map of the electron density in the Li_2 molecule.

nuclear attractors are not an artefact owing to approximations made when generating the wave functions (Edgecombe and Smith, 1993). They also occur in small sodium clusters and even solids. The electronic charge in the regions dominated by the non-nuclear attractors is loosely bound and delocalized. It is speculated that the non-nuclear attractors play an important role in the binding of metals and in their conducting properties.

Summary

- One of the strengths of science is that it prefers theories with economy of principle and assumption. Following this prescription if we can, we propose to study the total electron density using itself as a reference density rather than invoking an arbitrary and artificial reference density. This 'internal comparison' or 'introspection' is enabled by the gradient of the electron density.
- The gradient leads to the gradient vector field via the concept of a gradient path. The gradient vector field is the basis for the partitioning of a molecular system into mutually exclusive (or disjoint) regions of space which we will identify later as *atoms*. It is the practical foundation of the theory of atoms in molecules.
- Gradient paths occur in many familiar physical situations. They are orthogonal to envelopes for which the scalar function is constant. Gradient paths have a sense: they originate at one point and terminate at another. They never cross but they can meet at a point where

$\nabla\rho = \mathbf{0}$. Such a point is called a critical point (CP). The combination of the last two properties implies that gradient paths can be attracted to one type of critical point.

- Nuclei almost exactly coincide with this type of CP but it is possible that gradient paths are attracted to a non-nuclear attractor (NNA). In most cases the nucleus practically coincides with the position of the attractor and is called a nuclear attractor (NA). The (N)NA dominates a region uniquely traversed by paths terminating at the attractor.
- The gradient vector field naturally partitions a wide variety of systems into such regions. This analysis and everything that follows from it in subsequent chapters applies across chemical disciplines.
- There is a second kind of critical point, typically between nuclei, which will be classified later.

References

Alonso M. and Finn E. (1983). *Fundamental University Physics* Vol. 2, 2nd edn. Wokingham: Addison-Wesley.

Edgecombe K. E. and Smith V. (1993). Nonnuclear maxima in the charge density. *Zeitschrift für Naturforschung*, **48a**, 127–133.

Gatti C., Fantucci P. and Pacchioni G. (1987). Charge density topological study of bonding in lithium clusters. Part I: Planar Li_n clusters ($n = 4, 5, 6$). *Theoretica Chimica Acta*, **72**, 433–458.

Tsirelson V. G., Zou P. F., Tang T. H. and Bader R. F. W. (1995). Topological definition of crystal structure: determination of the bonded interactions in solid molecular chlorine. *Acta Crystallographica*, **A51**, 143.

Further reading

Marsden J. E. and Tromba A. J. (1981). *Vector Calculus* 2nd edn. San Francisco: Freeman.

Spiegel M. R. (1974). *Schaum's Outline of Theory and Problems of Advanced Calculus*. New York: McGraw-Hill.

Chapter 3

The atom

Can one cut a molecule
Without cutting one's fingers as well?

3.1 The atomic basin

In Chapter 2 we noticed that the nucleus is an attractor for a multitude of gradient paths. These paths originate at infinity and terminate at the nucleus after traversing a section of space known as the *atomic basin*. An *atom* is then defined as the union of the attractor and its associated atomic basin. 'The regions of space dominated by a nucleus', as we called them previously, are in fact the atoms of the theory of atoms in molecules. We recollect that the gradient vector field *naturally* partitions the molecule into atoms, i.e. the gradient of ρ carves the atoms from the molecule or system to which they belong. In that sense we can designate these atoms by the term *molecular atoms* as opposed to free or isolated atoms. Thus, every molecule falls apart into non-overlapping molecular atoms as shown on the front cover of this book. An aggregate of molecules also falls apart into (molecular) atoms, determined solely by the shape of the gradient vector field.

Every type of nucleus appears inside thousands of possible molecular atoms. In fact, there are millions of carbon (molecular) atoms because each atom is cut out of a particular chemical molecular environment of which there are as many as there are molecules. In a manner of speaking, every molecular atom is endowed with properties it inherits from the molecule of which it is a part. In other words, the atom reflects the features of its particular chemical environment.

Figure 3.1 shows a 3D representation of the oxygen atom in methanal as an example of an atomic basin. We recall that every gradient path crosses the constant electron density envelopes in an orthogonal way although the envelopes are not displayed here. This figure should be compared with Figure 2.6 where we displayed a 2D intersection of the gradient vector field with the molecular symmetry plane.

It emerges that large portions of space are not part of a given atomic basin. Put differently, gradient paths attracted to a given nucleus never reach the parts of space that are dominated by other nuclei. As a result, atomic

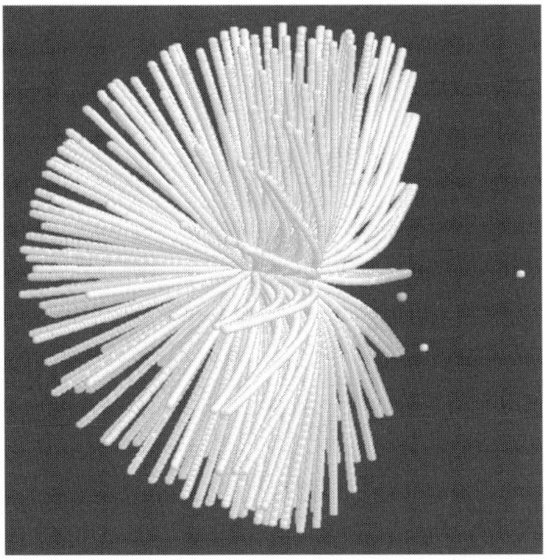

Figure 3.1 A 3D representation of gradient paths traversing the atomic basin of the oxygen atom in methanal. The small grey sphere just right of the bundle of gradient paths is the carbon nucleus and the two white spheres to the right of it are the hydrogen nuclei. The oxygen nucleus is hidden in the middle of the bundle. Note the conical tube of gradient paths, centred at the oxygen nucleus and protruding towards the carbon nucleus.

basins and thus atoms have sharp boundaries inside the molecule they belong to. Of course, they can extend to infinity at other sides where they do not neighbour another atom. For example, the oxygen atom in Figure 3.1 avoids the blank space at the right of the picture because this belongs to the neighbouring carbon atom. On the other hand, this oxygen extends to infinity at the left side. In general, an atom is open-cnded in some directions and bounded in others. Readers interested in the mathematical details of describing the inner shape of an atom should consult Box 3.1.

Box 3.1 A closer look

Natural atomic coordinates

In order to obtain a simple description of a system or an object, one frequently introduces special coordinates reflecting the intrinsic nature of that object. For example, an object with near spherical symmetry is best described using spherical polar coordinates. An axially symmetric system will benefit from the use of cylindrical coordinates. In similar vein, the atom has its own convenient set of coordinates called *natural atomic coordinates*. These coordinates resemble the familiar spherical polar coordinates (r, θ, φ) because it is appropriate to describe the atom from the point of view of the nucleus.

Indeed, the angular coordinates θ, φ constitute two of the three natural atomic coordinates and they are defined in the usual way (see Figure 1). The radial coordinate r, however, is replaced by the path coordinate s which varies as one moves along a gradient path. Figure 1 clarifies this set of natural atomic coordinates denoted by (s, θ, φ).

How do we interpret the natural atomic coordinates? Select an initial point on a sphere of arbitrary radius centred on the nucleus dominating the atomic basin. This initial point has the natural atomic coordinates $(s_0, \theta_0, \varphi_0)$. We choose to set the value of s_0 to zero to label the initial point. Increasing the value of the path coordinate s extends the gradient path from the initial point towards the nucleus where s attains the value $+\infty$. Decreasing s to negative values traces the path in the opposite direction, away from the initial point, towards infinity. At the origin of the gradient path s adopts the value $-\infty$.

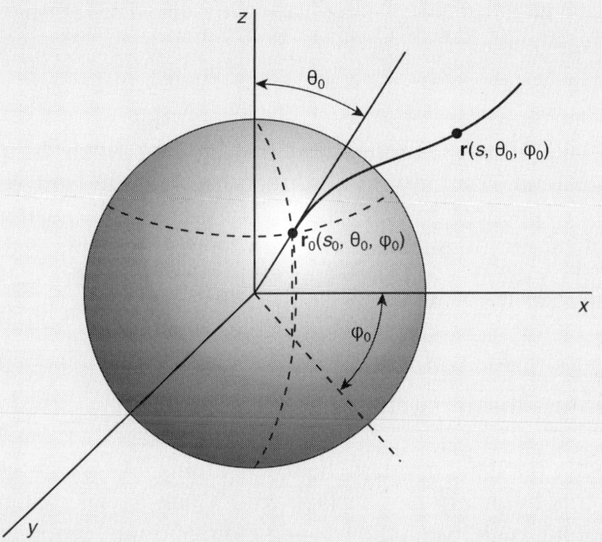

Figure 1 The natural atomic coordinates (s, θ, φ) with respect to a nucleus positioned at the origin of the (x, y, z) coordinate system. The angular coordinates θ and φ are defined in the same way as the corresponding spherical polar coordinates. One gradient path is shown in bold passing through an initial point with position vector $\mathbf{r}_0$. A change in the path coordinate s yields a point with position vector $\mathbf{r}$ lying further along the gradient path.

The path coordinate s is related to the actual gradient path length l via the equation:

$$ds = \frac{dl}{|\nabla\rho|} \tag{1}$$

From this equation it is clear that s reaches infinite values if $|\nabla\rho| = 0$, which happens at the origin and terminus of the gradient path. Using equation (1) we modify the equation in Box 2.2 to obtain:

$$\frac{d\mathbf{r}}{ds} = \nabla\rho \tag{2}$$

The solutions of this deceptively simple set of ordinary differential equations are the trajectories of the gradient paths (or integral curves) given by:

$$\mathbf{r}(s) = \mathbf{r}(s_0) + \int_{s_0}^{s} \nabla\rho(\mathbf{r}(t))dt \tag{3}$$

This equation demonstrates that a gradient path is in fact a collection of position vectors $\mathbf{r}$ determined by systematically varying the parameter (or coordinate) s over its entire range. Clearly, when $s = s_0$ we recover the initial point which determines the whole path. Obtaining practical solutions for equation (2) is non-trivial and computationally demanding, especially when $|\nabla\rho|$ is very small.

Finally it should be mentioned that free atoms (which do not have any neighbours) extend to infinity in all directions. The gradient vector field of a free atom in a spherical ground state is very simple: it is a pattern of straight lines radiating from the nucleus. This can be readily explained because we know that the constant ρ envelopes of an isolated atom are spheres and the gradient paths are everywhere perpendicular to these envelopes.

3.2 The interatomic surface

We now focus on the boundary between atoms. A glance at Figure 3.2 reveals that this boundary is a surface with a non-trivial shape. This figure shows the boundary between the oxygen atom and the carbon atom again in methanal. It is clear from this figure that the interatomic boundary consists of a bundle of gradient paths originating at infinity and terminating at the critical point (CP) between the nuclei. This topological object is called the *interatomic surface* (IAS). As expected for a boundary such as the IAS, it does not contain any gradient paths which are attracted to either nucleus. Strictly speaking, the IAS is a truly 'independent' object in that it lies perfectly between two atoms and is not part of either. The IAS between two neighbouring atoms A and B is denoted by the symbol S_{AB}.

The IAS is sometimes called a *zero-flux surface* (details, see Box 3.2). To understand this name we consult Figure 3.3 where the gradient vector field of the diatomic carbon monoxide is shown. In this picture we analyse the difference between an arbitrary boundary surface between carbon and oxygen (labelled **2**) and the actual IAS (**1**). We select an arbitrary point on both

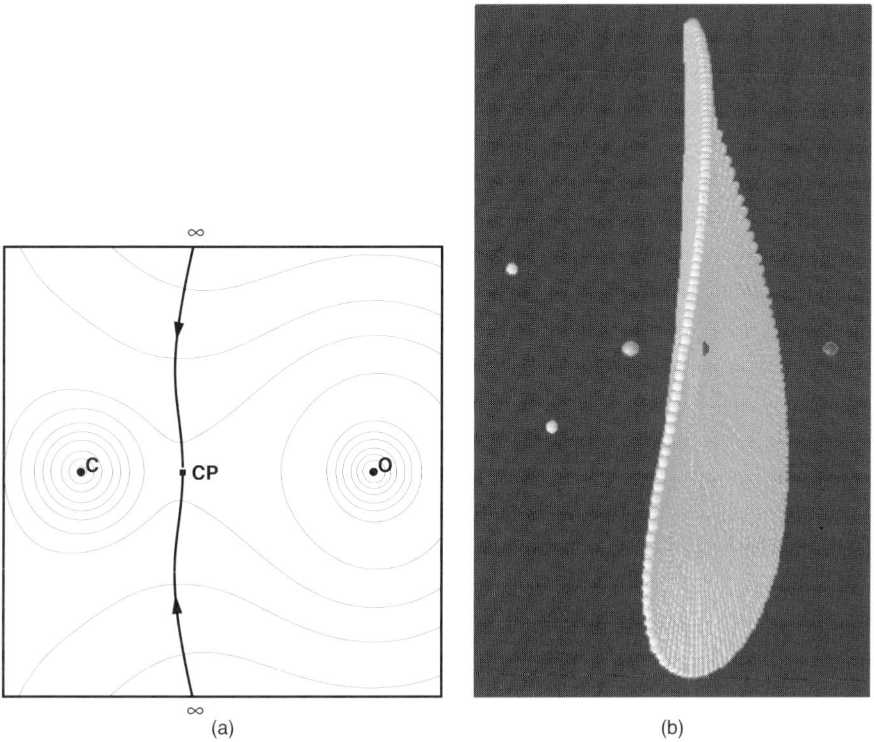

Figure 3.2 (a) The intersection of the interatomic boundary with the molecular plane superimposed on the contour map of the electron density. The CP at the centre of the boundary is labelled by a square. The arrows indicate that the gradient paths constituting the boundary originate at infinity and terminate at the CP in between the nuclei. (b) A 3D representation of the boundary between the oxygen atom and the carbon atom in methanal. The CP in the centre of this boundary is marked by a dark grey sphere. The two hydrogen nuclei appear as white spheres on the right.

surfaces and measure the angle between the normal to the surface **n** and $\nabla\rho$ at each point. The IAS distinguishes itself from any other arbitrary surface in that at *every* point the vector **n** is orthogonal to $\nabla\rho$ or $\mathbf{n} \cdot \nabla\rho = 0$ *for all* points on the surface. In general, the flux of a vector field **F** through a surface patch is defined as the surface integral of $\mathbf{n} \cdot \mathbf{F}$ over the patch. The integrand of the flux of $\nabla\rho$ is $\mathbf{n} \cdot \nabla\rho$. It vanishes at *any point* on the IAS. As a result, the flux of $\nabla\rho$ through the whole IAS vanishes as well, such that the alternative name of zero-flux surface is justified.

With the aid of a branch of mathematics called *differential geometry* it is possible to characterize the intrinsic geometry of surfaces and other topological objects or *manifolds*. Differential geometry is the study of geometric figures using the methods of calculus and has been applied to IASs (Popelier, 1994, 1996a). One of the powerful quantities that differential geometry

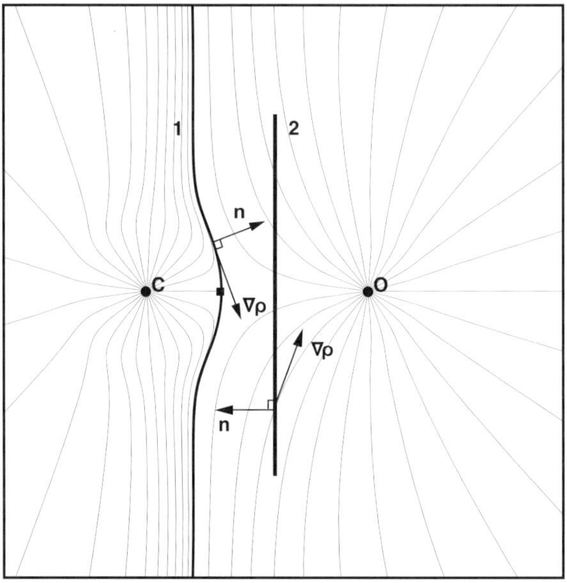

Figure 3.3 A simple representation of the gradient vector field in carbon monoxide focusing on the difference between the IAS (**1**) and an arbitrary surface boundary (**2**). At any point of the IAS, the normal to the IAS is perpendicular to $\nabla\rho$. This property is not shared by the arbitrary surface.

Box 3.2 A closer look

Zero-flux surfaces

Consider Figure 3.2(a) where the positive branches of a sequence of hyperbolae are shown to explain a general feature of zero-flux surfaces. This sequence of hyperbolae is described parametrically by the equation $xy = a$ where a decreases from 10 to 0. The smaller the parameter a, the closer the branch comes to the semi-axes of x and y. In the limiting or *degenerate* case of $xy = 0$ $(a = 0)$, the hyperbolic branch coincides with the x and y axes.

We now draw a parallel between this sequence of hyperbolic branches and bundles of gradient paths. Figure 3.2(b) shows the case of the diatomic carbon monoxide because the cylindrical symmetry simplifies our comparison. Each gradient path of this picture should in fact be convoluted around the molecular axis to form a bundle of gradient paths that it actually represents. This convolution turns the path originating at infinity and terminating at the CP between the nuclei (labeled 4′) into the IAS or zero-flux surface.

The key point is now that the sequence of gradient paths, labelled 1, 2, 3, 4 and 4′ in (b) looks like the sequence of hyperbolic branches in (a). As a result, the set of gradient paths (4, 4′) can be considered as a limiting case.

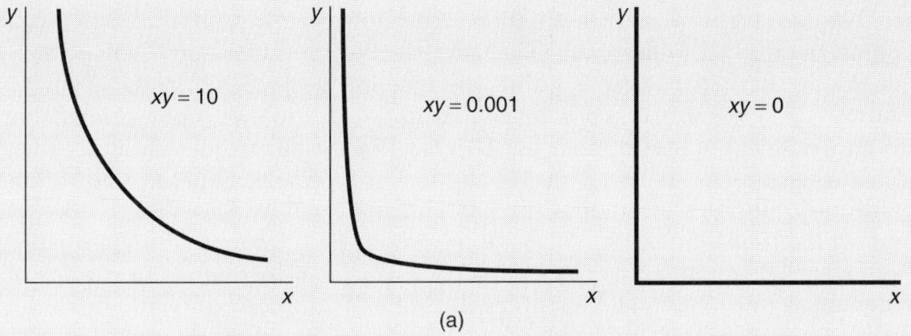

(a)

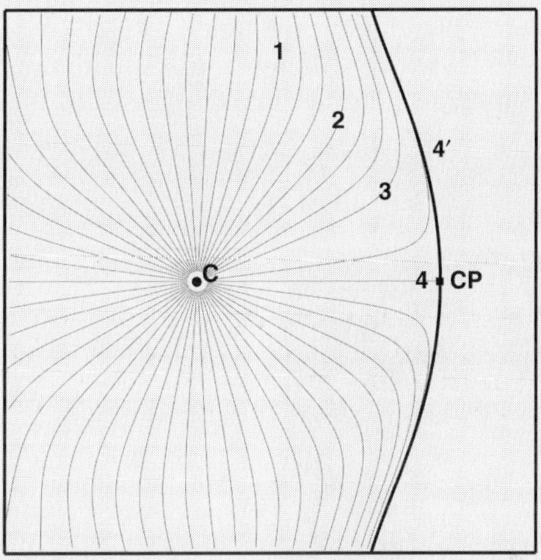

(b)

Figure 1 (a) A sequence of hyperbolic branches parametrically described by the equation $xy = a$. Note that the smaller a, the larger the curvature near the origin. When $xy = 0$ the hyperbolic branch coincides with the two semi-axes x and y. This is a degenerate situation which corresponds to the set of gradient paths labelled 4, shown in Figure 1(b). (b) A blow-up of the gradient vector field of the carbon atom in carbon monoxide. The square marks the CP between the nuclei, and the bold line represent the interatomic surface S_{CO}. The set of increasing numerical labels refers to gradient paths that show an increasingly pronounced curvature near the CP. The gradient path labelled 4 is in fact the limiting case of the sequence of paths. It consists of two separate gradient paths: one originating at the CP and moving towards the nucleus; the other originating at infinity and terminating at the CP.

Of course, the correspondence with the equation $xy = 0$ should not be taken too literally, but the idea of degeneracy is definitely transferable.

The insight gained in Figure 1(b) serves to highlight a subtle but important point about the mathematical definition of a zero-flux surface or IAS. The equation for an IAS is $\mathbf{n} \cdot \nabla\rho(\mathbf{r}) = 0$ for all points on the surface, or in set-theoretical short-hand: $\forall \mathbf{r} \in S$ where $\mathbf{r}$ is the position vector corresponding to the point. What does this equation mean in words? It states that, at a given point, the normal to the surface is orthogonal to $\nabla\rho$. This is an involved way of saying that, at that given point, the gradient vector $\nabla\rho$ is *parallel* to the surface. In other words, $\nabla\rho$ is tangential to the IAS in that point. Surely this is equivalent to asserting that the IAS is in fact a bundle of gradient paths. Then the question is, what makes the IAS so special compared with other (arbitrary) bundles of gradient paths? The answer is that $\mathbf{n} \cdot \nabla\rho(\mathbf{r}) = 0$ holds *for all the points on the IAS.* Take the gradient path labelled 3 in Figure 1(b) and rotate this path about the symmetry axis of the diatomic. The *surface of revolution* thus obtained constitutes a bundle of gradient paths for which the equation $\mathbf{n} \cdot \nabla\rho(\mathbf{r}) = 0$ holds except in one point: *the nucleus.* Indeed, at the nucleus the gradient of ρ is not defined because ρ shows a cusp here. This means that $\nabla\rho$ is *not* defined and therefore $\mathbf{n} \cdot \nabla\rho(\mathbf{r}) = 0$ does *not* hold at the nucleus. By virtue of this only difference the equation $\mathbf{n} \cdot \nabla\rho(\mathbf{r}) = 0, \forall \mathbf{r} \in S$ singles out the IAS from an infinite number of zero-flux surfaces that contain the nucleus.

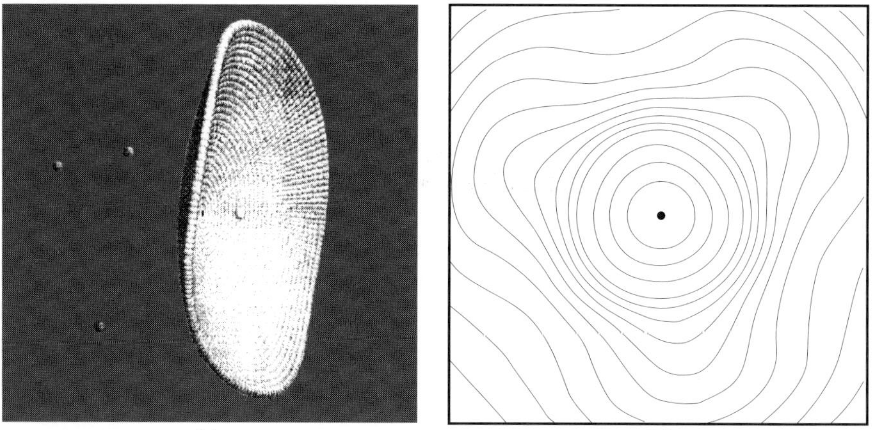

(a) (b)

Figure 3.4 (a) A 3D representation of the C | H interatomic surface in methane. This picture should be compared with the corresponding contour map of Figure 3.4(b). (b) A contour map of the Gaussian curvature K of the IAS between C and H in methane. The outer contour line corresponds to a curvature of 0.001 and the contours increase as 0.005, 0.01, 0.015, 0.02, 0.025, 0.03, 0.035, 0.04 0.045, 0.05, 0.06, 0.07, 0.08 and 0.09. The CP in the centre of the IAS is marked by a dot. Note that the expected threefold symmetry becomes more pronounced towards the outer regions of the IASs. (From Popelier, 1996a. Reproduced by permission.)

defines is Gaussian curvature, which, omitting mathematical details, is an invariant property independent of the surface's representation and measures its local curvature. It is derived from the *fundamental forms* of the surface which completely determine the surface in the same way that curvature and torsion determine a curve in real space. Figure 3.4 shows one of the four

IASs between C and H in methane, denoted C|H, and a contour diagram of its Gaussian curvature. The figure demonstrates that IASs possess more internal structure than is apparent by mere inspection. The *total curvature* is defined as the surface integral of the Gaussian curvature of a surface patch. For example, the total curvature of a sphere is 4π and that of a plane or torus is 0.

Further differential geometrical research into AIM manifolds is warranted, but some encouraging results have already been obtained. For example, the total curvature of an IAS measures the chemical distortion of an atom in a molecule, which is caused by a combination of polarization deformation, charge transfer and steric effects.

3.3 The shape of an atom

Given a molecule and an arbitrary point in space, how can we determine to which (molecular) atom this point belongs? Figure 3.1 suggests the obvious solution of tracking down the terminus of the gradient path passing through the given point. We thereby recover the attractor or nucleus that dominates the atomic basin. Although this procedure is computationally cumbersome, it is a conceptually simple way to find the set of points that constitute an atom (Popelier, 1998). But we should focus not just on the termini of the gradient paths but also upon how they 'propagate' through space. Their flow pattern (or gradient vector field) reveals the inner structure of the atom, i.e. the way space itself seems to be warped or distorted around the nucleus. At the moment, however, it is not clear how literally this comparison can be taken.

Although no formal comparison has been proposed, it is tempting to think of the atom as a little universe in the sense of the theory of general relativity, loosely speaking. Anyone 'travelling' inside the atom, using natural atomic coordinates (see Box 3.1), will have the impression that the atom is infinite. A more careful investigation of the atom's inner curvature, however, should reveal that the atom is bounded in certain regions. Again, differential geometry is the ideal tool to study the intrinsic geometric structure of the atom's space in the hope of linking it to chemical concepts. Despite this intriguing proposition, in the following we will focus only on the 'outer' shape of the atom, i.e. the atom will be regarded as a subspace bounded by surfaces.

An atom, which is a fragment of an isolated molecule, typically has considerable 'open' parts that extend to infinity. So the atom is bounded inside the molecule by IASs but it is open or unbounded at the exterior of the molecule. Inspired by the envelope representation in Figure 1.2, we propose to 'cap' the atom at the open end by the 0.001 au envelope constant electron density of the entire molecule. This choice is not a fundamental or rigorously justifiable conceptual contribution of AIM but is based simply on an adequate representation of the van der Waals envelope of a molecule (Bader *et al.*, 1987). As expected, the atoms are also capped by this 0.001 au edge of the molecule. Figure 3.5 shows examples of atomic shapes drawn from

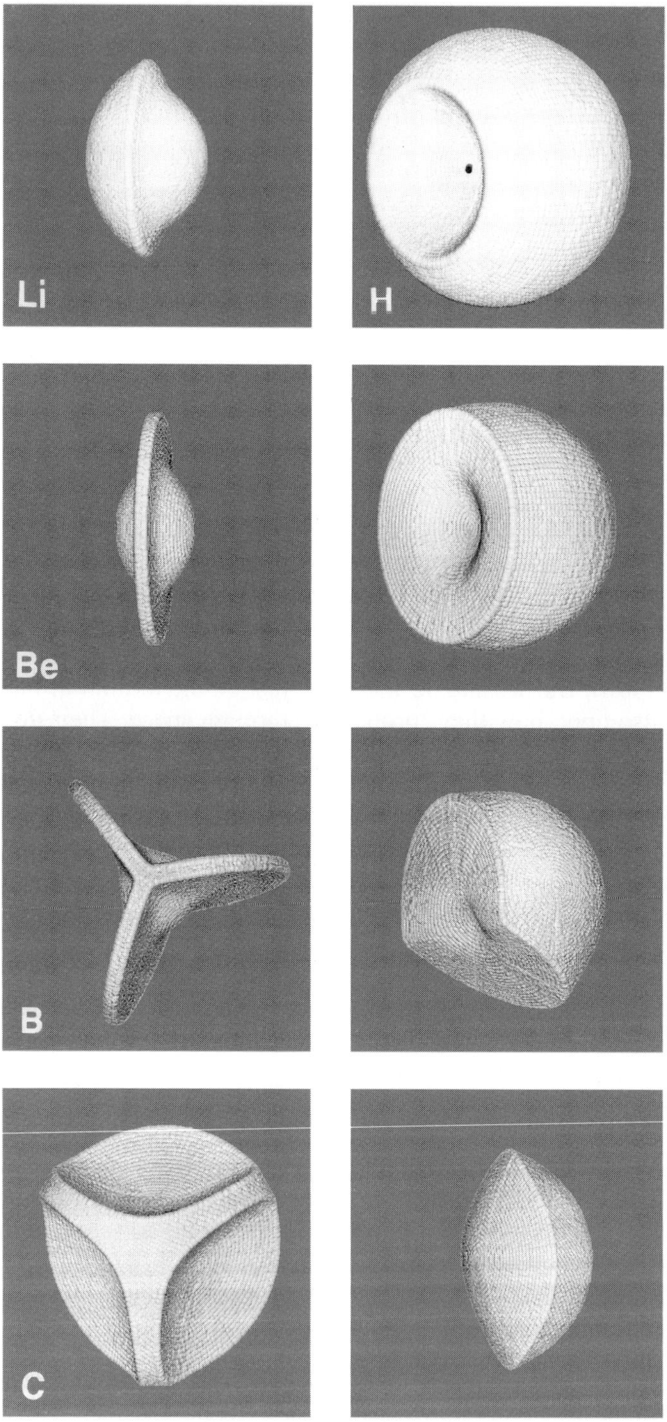
Li
H
Be
B
C

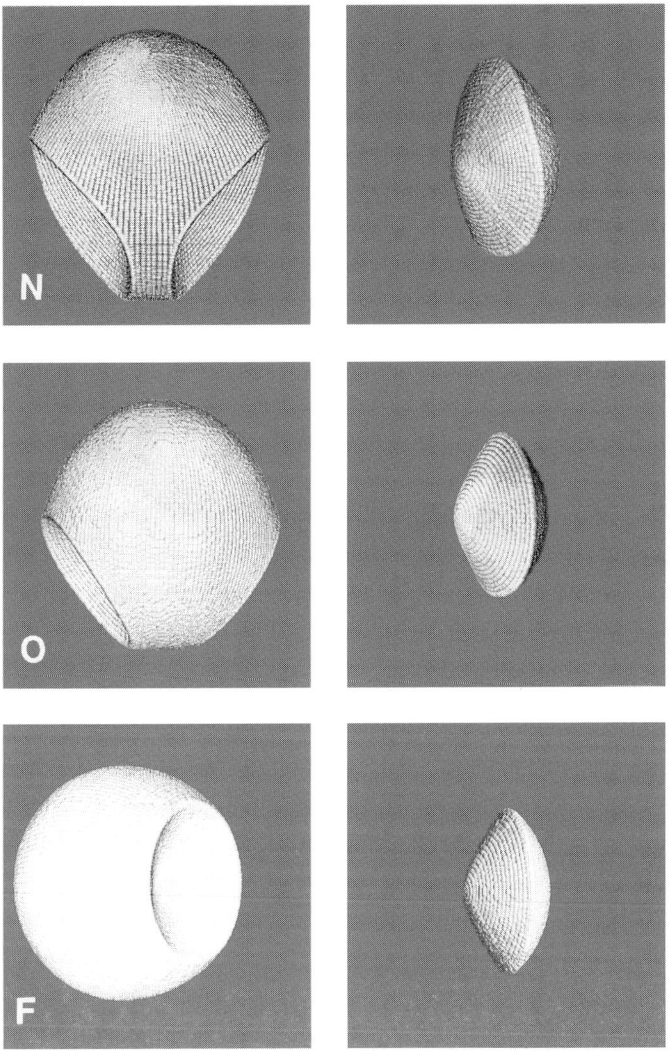

Figure 3.5 Representations of the atoms in the second-row hydrides AH_n: LiH, BeH_2, BH_3, CH_4, NH_3, H_2O and HF. The surfaces bounding the atoms consist of two types: interatomic surfaces and patches of the $\rho = 0.001$ au molecular envelope. (From Bader *et al.*, 1994. Reproduced by permission.)

the second-row hydrides AH_n. It is clear that only that part of the IAS *inside* the $\rho = 0.001$ au envelope preserves the atom's boundary inside the molecule. The remainder of the IAS, which extends to infinity, is ignored in this practical definition of an atom. In summary, the atomic boundary consists of the union of IAS patches (inside a constant ρ envelope) and parts of the constant ρ envelope.

In contradistinction to most chemists, theoretical chemists seem not to think of atoms as finite, almost tangible objects in real space. Although we are aware of the practical advantages of alternative definitions of atoms, one should contemplate the full chemical picture that AIM offers and remember also that it is anchored in quantum mechanics. Perhaps the most remarkable property of AIM's atoms is the fact that one can rigorously define the energy of an atom *inside* a molecule (see Chapter 7). This paramount property is not based on the features of any practical computational scheme but is deeply ingrained in quantum mechanics itself.

The problem of carefully defining the edge of a molecule is a complicated matter that would justify a chapter by itself, but within the context of AIM it suffices to make a few comments. The idea of representing the spatial shape and extent of a system by centring 'atomic' spheres of fixed radii at nuclei goes back to Bragg, Pauling, Slater and others. Although originating in ionic crystals, this idea has been extended to free neutral atoms and even to molecules. In the late 1960s it was realized that the $\rho = 0.002$ au envelope was a useful measure of molecular size and packing in crystals. On the other hand, the $\rho = 0.001$ au envelope turned out to agree better with molecular volumes in the gas phase, as determined by the second virial coefficient or viscosity data fitted with a Lennard–Jones potential. Note that either choice (0.001 or 0.002 au) encloses almost the entire electronic charge (98–99%) of the atom. How this can be computed is explained in the next section.

3.4 Atomic integration

From now on we denote the atomic basin or the portion of space the atom occupies by Ω. *Atomic properties* are then defined as volume integrals over Ω of some integrand which we have to specify for each property. Many properties are hard to define without quantum mechanical considerations (Chapter 7), but at this stage it is straightforward to define two atomic properties: the *atomic volume*, $v(\Omega)$ and the atom's *electronic population*, $N(\Omega)$. The latter is simply defined as:

$$N(\Omega) = \int_{\Omega} d\tau \rho(\mathbf{r})$$

and the former as:

$$v(\Omega) = \int_{\Omega} d\tau$$

where the integrand is trivially unity. Note that the integration volume of the population integral extends far beyond the $\rho = 0.001$ au envelope because ideally it should sample all electronic charge, up to infinity. Since all integrals over atoms are computed numerically we cannot really integrate to infinity but only to a practical boundary which can be taken as infinity. Since the electron density drops to zero fairly rapidly, such a practical boundary is the $\rho = 10^{-6}$ au envelope. The integration range corresponds to a distance between nucleus and outer boundary of the order of 10 Bohr, depending on the atomic number.

No matter how trivial the volume integral $v(\Omega)$ looks, it is a very demanding task to design a robust and efficient algorithm to perform an integration over an arbitrary atomic shape. The volume integral can be expressed in terms of two plausible coordinate systems: natural atomic coordinates or spherical polar coordinates centred at the nucleus. The former choice of coordinates leads to an elegant but impractical integration scheme (Biegler-König *et al.*, 1981) whereas the latter choice is the one implemented in all current AIM programs (Biegler-König *et al.*, 1982; Stefanov and Ciosłowski, 1995; Popelier, 1996b). Since analytical integration (in either coordinate system) is forbidding, a numerical quadrature in three dimensions is invoked. In essence, a quadrature approximates the integral by a sum consisting of terms which are products of a particular weight and the integrand evaluated at predetermined grid points. Each angular grid point, i.e. (θ_i, φ_i) value (see Box 3.1) corresponds to an *integration ray* centred at the nucleus. The 3D quadrature consists of summing over all angular grid points and over the radial grid installed on each integration ray.

Determining the length of each integration ray as a function of (θ, φ) is the most challenging part of the integration algorithm. Imagine the integration ray 'sweeping around' the atomic basin and probing the boundaries of the atom or arbitrary shape. Finding the intersection of the integration ray with an atom's boundary (either an IAS or a constant electron density envelope) needs careful thought because of multiple intersections and poorly described IASs. Obtaining atomic properties is still a time-consuming activity in spite of many improvements of algorithm. This is probably one of the reasons that these calculations are still not routinely performed in wider circles of the chemical community. However, at the time of writing, integrations can be routinely performed on personal computers (PCs), although an acceleration of current algorithm is desirable.

Table 3.1 furnishes some numbers illustrating typical atomic volumes and populations as found in the equilibrium geometry of methanal. Although the quality of integration of a single atom can be measured via an atomic quantity (see section 7.1), a comparison between the total molecular electron population and the sum of all atomic populations can be used as an estimate as well. It is clear from Table 3.1 that these two values tally to at least four decimal places. The number of quadrature points needed to produce a reliable atomic property depends on the property and on the shape of the atom. Atoms with long tails and a cusp-like shape, such as those occurring in ring systems (see Chapter 5), require large quadrature grids. Furthermore, the atomic volume is more sensitive than the population to the quality of the grid.

Table 3.1 The volumes and populations of the atoms in methanal (in au).

Atom	$v(\Omega)$	$N(\Omega)$
C	66.39	5.0188
O	138.36	9.0601
H	50.48	0.9605
Total	**305.71**	**16.0000**

3.5 Atomic additivity

An important advantage of finite atoms is that they do not overlap. Each atom has a sharp and well defined boundary inside the molecule given by the set of bounding IASs. In other words, there are no gaps between atoms carved out of a given molecule: they perfectly match and compose the original molecule like the pieces of a jigsaw puzzle. In this sense the atoms literally add up to form the molecule they are part of, hence the attribute of *additivity*. The power of this attribute is that not just shape and volume are additive but also any physical property that belongs to the atom. This statement is expressed simply via the following equation:

$$A_{\text{molecule}} = \sum_{\Omega} A(\Omega) \tag{3.1}$$

where A represents a property such as energy, a population or volume.

Two consequences of this attribute of additivity are the appearance of symmetry in the atoms and the construction of functional groups. If the nucleus of an atom coincides with the centre of molecular symmetry then the atom adopts the point group of the whole molecule. For example, the shape of the carbon in methane (Figure 3.5) has tetrahedral symmetry. This atomic symmetry is also reflected in properties other than shape. For example, all components of the atomic quadrupole moment of carbon in methane vanish. If the nucleus of an atom does not coincide with the centre of molecular symmetry the atom will not adopt all the symmetry features of the molecule. However, interesting symmetry relationships between atoms may emerge; for example, the two hydrogen atoms of methanal are mirror images of each other.

Functional groups are defined in a straightforward manner owing to atomic additivity: they are simply the union of the constituting atoms. For example, the carboxyl group is just a set of four atoms: carbon, hydrogen and two oxygen atoms. Of course, these atoms have to be sampled from the appropriate carboxyl fragment as it occurs in a molecule. Figure 3.6 further illustrates this idea showing a 3D picture of the glycyl group which we denote by the shorthand $|NH–C_{\alpha}H_2–C(=O)|$ where the symbol '|' represents an IAS as before. This functional group consists of seven atoms and represents the

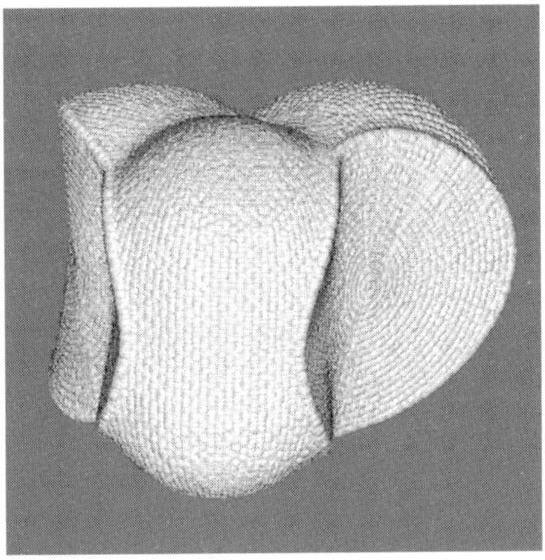

Figure 3.6 Representation of the glycyl group $|NHCH_2C(=O)|$ with the $-|NH$ IAS on the left and the $C(=O)|-$ IAS on the right. The remainder of the group is represented by the 0.001 au envelope of electron density with the CH_2 in the foreground. This picture gives an example of seven atoms forming a functional group. (From Bader and Popelier, 1993. Reproduced by permission.)

glycine amino residue in its peptidic environment. It is a neutral fragment (i.e. no net charge) that can be studied in its own right as a recurring unit with useful chemical properties (Bader *et al.*, 1992). A question that naturally arises is, how transferable are the properties of such a unit?

3.6 Atomic transferability

There are literally many millions of molecular atoms because there are millions of molecules which all give rise to a set of constituent atoms. Nevertheless the sometimes bewildering shapes of atoms have been criticized as being contrary to chemical intuition. This should not be disconcerting, rather it could be interpreted as an expression of the richness of chemistry. Indeed, the amazing variety of atoms is a result of quantum systems cutting themselves into fragments, each leaving behind on the fragments detailed fingerprints of the total molecule. Is it possible, then, to find exactly the same atom more than once coming from different molecules?

This phenomenon would be equivalent to perfect transferability, which has been proven to be an unattainable limit. Disregarding mathematical details there is a basic theorem that says that if two continuous analytical functions $f(x)$ and $g(x)$ are identical over a given interval $[a, b]$ then they must be equal over the whole real axis. Again omitting detailed considerations (Bader and Becke, 1988), we obtain as a corollary of this theorem that if ρ is equal over the Ω spaces of two atoms A and B, i.e. $\rho_A(\mathbf{r}; \Omega) = \rho_B(\mathbf{r}; \Omega)$, then ρ must be equal over whole space or $\rho_A(\mathbf{r}) = \rho_B(\mathbf{r})$. This implies that these two supposedly different atoms *A* and *B* were actually coming from the same molecule. Consequently it is not possible to cut an atom from one molecule and insert it *exactly* into a corresponding cavity of another molecule. Thus it is impossible to transfer perfectly an atom from one molecule to a different molecule.

Transferability is possible to some extent. The 'zoo' of atoms does reveal subsets containing atoms with striking patterns of mutual similarity that can be analysed using mathematical techniques collectively known as *cluster analysis*. It should be emphasized, though, that transferability is something one *observes* in chemistry rather than *imposes* on it. Supported by the faith that a method reveals the correct degrees of transferability, this information can be used to set up a library of atoms, enabling a rapid and accurate construction of large ensembles of atoms such as proteins. It is important to realize that transferability is a continuous rather than binary (yes/no) concept for it depends on what errors in information transfer are allowed.

Keeping with the analogy of the diversity of life forms (the zoo), it might be exciting to speculate on a theory of 'atomic classification and evolution', paralleling the principles of biological classification and evolution. However, on a less ambitious note we close this section with some of examples of transferability. Using AIM it is possible to reproduce the experimentally observed transferability of heats of formation of the methylene group ($|CH_2|$) in normal hydrocarbons. Successful transferability, however, not only is achieved by recovering well known observations of *n*-alkanes but also encourages the use of transferable biochemical fragments. For example, the dipeptide $H_2N{-}C_\alpha H_2{-}C(=O)|NH{-}C_\alpha H_2{-}COOH$ can be constructed from two straightforward fragments obtained by cleaving the central peptide bond along the C|N IAS. The left fragment can also be obtained by cleaving $H_2N{-}C_\alpha H_2{-}C(=O)|NH_2$ and the right fragment by cleaving $HC(=O)|NH{-}C_\alpha H_2{-}COOH$. The difference between the total energy of the original dipeptide and the sum of all atomic contributions to the two fragments amounts to only 12 kJ/mol (Chang and Bader, 1992). This is even more remarkable when one realizes that the energy of any non-hydrogen atom in the fragments is of the order of 10^5 kJ/mol.

Summary

- An atom is the union of an attractor (the nucleus) and its associated atomic basin. It has an inner structure expressed by the intrinsic geometry of the gradient vector field.
- An interatomic surface or zero-flux surface is a topological manifold which separates atoms. It consists of a bundle of gradient paths terminating at the CP between two atoms.

- There is an enormous number of atomic shapes bounded by a combination of IASs and constant electron density envelopes.
- Simple atomic properties (for example, volume and electronic population) can be expressed as volume integrals over an atomic basin. The quadrature scheme that performs such integration is computationally demanding and very complicated.
- As atoms are finite, non-overlapping entities leaving no gaps and they exhaust space. Consequently their properties can simply be added to obtain the properties of a functional group.
- Perfect transferability of atoms from one system to another is an unattainable limit but, if observed and present, a large degree of transferability will faithfully be revealed by AIM.

References

Bader R. F. W. and Becke P. (1988). Transferability of atomic properties and the theorem of Hohenberg and Kohn. *Chemical Physics Letters*, **148**, 452–458.

Bader R. F. W. and Popelier P. L. A. (1993). Atomic theorems. *International Journal of Quantum Chemistry*, **45**, 189–207.

Bader R. F. W., Carroll M. T., Cheeseman J. R. and Chang C. (1987). Properties of atoms in molecules: atomic volumes. *Journal of the American Chemical Society*, **109**, 7968–7979.

Bader R. F. W., Popelier P. L. A. and Chang C. (1992). Similarity and complementarity in chemistry. *Theochem-Journal of Molecular Structure*, **87**, 145–171.

Bader R. F. W., Popelier P. L. A. and Keith T. A. (1994). Theoretical definition of a functional group and the molecular orbital paradigm. *Angewandte Chemie* (International edition), **33**, 620–631.

Biegler-König F. W., Nguyen-Dang T. T., Tal Y., Bader R. F. W. and Duke A. J. (1981). Calculation of the average properties of atoms in molecules. *Journal of Physics B: Atomic and Molecular Physics*, **14**, 2739–2751.

Biegler-König F. W., Bader R. F. W. and Tang T-H. (1982). Calculation of the average properties of atoms in molecules II. *Journal of Computational Chemistry*, **3**, 317–328.

Chang C. and Bader R. F. W. (1992). Theoretical construction of a polypeptide. *Journal of Physical Chemistry*, **96**, 1654–1662.

Popelier P. L. A. (1994). An analytical expression for interatomic surfaces in the theory of atoms in molecules. *Theoretica Chimica Acta*, **87**, 465–476.

Popelier P. L. A. (1996a). On the differential geometry of interatomic surfaces. *Canadian Journal of Chemistry*, **74**, 829–838.

Popelier P. L. A. (1996b). Integration of atoms in molecules: a critical examination. *Molecular Physics*, **87**, 1169–1187.

Popelier P. L. A. (1998). A method to integrate an atom in a molecule without explicit representation of the interatomic surface. *Computer Physics Communications*, **108**, 180–190.

Stefanov B. B. and Ciosłowski J. (1995). An efficient approach to calculation of zero-flux surfaces of atomic integration data. *Journal of Computational Chemistry*, **16**, 1394–1404.

Further reading

Struik D. J. (1961). *Differential Geometry*. Wokingham: Addison-Wesley.

Morgan F. (1993). *Riemannian Geometry*. Boston, MA: Jones & Bartlett.

Chapter 4

The bond

How do you do? My name is Bond,
Chemical Bond . . .

4.1 Origin of the bond concept

Anyone who takes a standard chemistry textbook off the shelf with the intention of finding *the* (or just *a*) definition of the chemical bond will soon be disappointed. Moreover, even specialized books dedicated to chemical bonding fail to supply a clear-cut definition of a bond. Yet most chemists seem to have strong opinions on what constitutes a bond and what does not. Why is this cornerstone of chemistry not defined? AIM, in fact, does provide a clear definition of a bond (see section 4.3). Moreover, AIM extracts this crucial information from solutions of the Schrödinger equation via the topology of the electron density. Therefore the definition that AIM proposes is, as usual, embedded in quantum mechanics and fully compatible with modern *ab initio* wave functions.

Let us step back for a moment and try to understand the origin of the chemical bond. The early concept of bonding is closely related to the concept of valence. In the middle of the nineteenth-century, chemists were aware of the striking regularities in the proportions that were found between certain elements with respect to other elements. Observations of many formulae of inorganic compounds revealed recurring chemical equivalent ratios. Although the roots of the notion of valence lay in ionic (inorganic) compounds it was later introduced in organic chemistry. Kekulé and Couper independently postulated that carbon was tetravalent and that carbons could form bonds with one another. This could be conveniently expressed by structural formulae with lines between 'linked' atoms. The original definition of *valence* states that this is the number of hydrogen atoms that will *combine* with one atom of the element. If we take for example methane, then what does 'combine' mean? Does it simply mean that carbon in methane always occurs together with four hydrogens? Does that imply that carbon *is bonded* to each of these four hydrogens? What gives us the right to focus all attention on carbon and ignore the relationship between the hydrogens? If chemists' opinions on what is bonded or not is based on these rather vague considerations of 'accompaniment' then it is not surprising that simple bonding schemes sometimes break down. Boranes, for example, have challenged conventional views on which atoms are bonded or not.

At this stage it is important to pause and scrutinize the idea of the bond in terms of the observations that the nineteenth-century pioneers of chemistry could make. The fact that certain atoms always occur in the company of one or a group of other atoms gave rise to valence. This connectivity was then naturally expressed in terms of special relationships or links between atoms, known as bonds. Thus in the early days of chemistry, valence and bonds went hand in hand, expressing two aspects of the same phenomenon: the occurrence of several atom types in regular ratios or patterns. This fixation with satisfying a constant valence for carbon led Kekulé to the useful idea of multiple bonds and a structural formula for benzene, but does not offer an explanation for hypervalent carbon. A hypervalent carbon is one to which more than four atoms (or groups) are bonded. We now prefer to call such a carbon hypercoordinated (Olah and Rasul, 1997). The original awkwardness surrounding hypervalency is the price one has to pay for squeezing chemistry into a theoretical framework based on 'well behaved' molecules. It is rewarding to abandon naïve pictures altogether and to propose a simple but rigorous bonding picture which can be supported by the Schrödinger equation.

The next development included an awareness of the geometry of the atoms participating in a bonding scheme. For example, it was realized that the four univalent hydrogens around the central carbon atom in methane occur in a tetrahedral arrangement. These advances marked the beginning of the common ball and stick model to represent a molecule as a set of atoms joined by bonds. The main question is, what gives us the right to join the carbon and a hydrogen nucleus as a bond in methane but not any two hydrogen nuclei? Although this question may seem purely academic, or even inappropriate, there are many familiar examples of molecules where it is *not* clear which nuclei should be joined to form bonds (for example, organolithiums, boranes, propellanes). If that question was ever raised by the nineteenth-century chemists they surely could not answer it by the admirable yet crude means that led them to the concept of a bond. The picture that Lewis and later Pauling offered, based on simple counting rules and the valence bond model, does not answer that question either. What these theories do provide is a host of molecular bonding schemes that chemists assimilate but perhaps do not challenge enough. Let us therefore review the problem of defining bonds without reference to early models.

4.2 Energy considerations

In this section we want to rephrase the bonding problem from the point of view of the energy of a molecule or molecular aggregate. The energy must

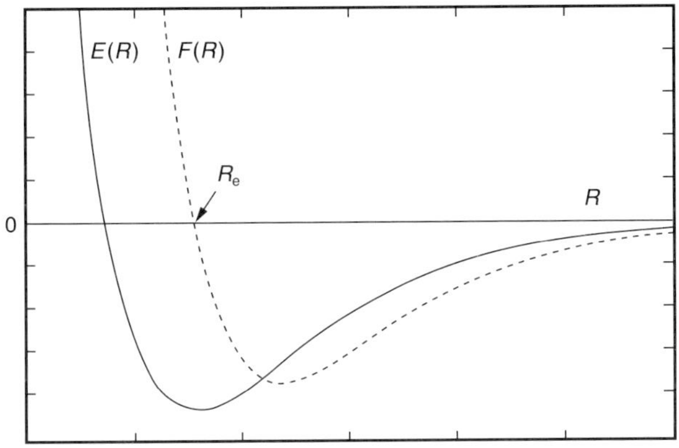

Figure 4.1 Plot in arbitrary units of the energy of a diatomic $E(R)$ (solid) versus the internuclear separation R, which is the potential energy curve for motion of the nuclei. Superimposed on this curve is the curve for $F(R)$ (dashed). The forces on the nuclei are attractive when $F(R) < 0$. The equilibrium separation R_e corresponds to the situation of vanishing forces.

be a good starting point for a fresh re-examination because it is a well defined quantity which can be measured and computed accurately. First we consider diatomic and then polyatomic molecules.

Within the Born–Oppenheimer approximation the (electronic) energy E of a diatomic AB can be completely described with one internal coordinate R which is the internuclear separation, i.e. the distance between the two nuclei A and B. Figure 4.1 shows a plot of the potential energy curve $E(R)$ for a typical diatomic. It is easy to prove that for any potential function the force on nucleus A, $\mathbf{F}_A$ is opposite in sign to the force $\mathbf{F}_B$ on nucleus B, or $\mathbf{F}_A = -\mathbf{F}_B$. Since these forces are directed along the internuclear axis we can replace the vector by a scalar quantity, i.e. one vector component is sufficient to describe the force. It is even more convenient to introduce one single force with respect to R, denoted by $F(R)$ and defined by $-dE(R)/dR$. This *internal* force is related to the *cartesian* forces on the individual nuclei via $F_A = -F_B = F(R)$. There is a special internuclear separation R_e where $F(R)$ vanishes. This corresponds to the equilibrium geometry and constitutes the lowest energy the molecule can attain.

The minimum in the energy curve of Figure 4.1 indicates that it is energetically more favourable for the separate atoms A and B to reside close to each other, at the equilibrium distance. Moreover, this bound state is stable because any compression or elongation of the internuclear separation R will lead to an energy increase. This is equivalent to stating that $F(R)$ is a force that restores the equilibrium situation. These features can be summarized by designating the potential energy profile as being *attractive* or *restoring*. This is expressed mathematically simply by demanding that the second derivative of the energy with respect to internuclear separation is positive, that is

$d^2E(R)/dR^2 > 0$. The word 'attractive' should not be confused with *attracting* (if $dE(R)/dR > 0$ or $F(R) < 0$) which qualifies the *first* rather than the second derivative of $E(R)$ with respect to R. This distinction is not universally made, but turned out to be crucial to avoid typical misconceptions in discussions about the essence of chemical bonding (see section 4.4).

Now, if one atom of the diatomic molecule were to be translated in space, the other atom would follow to preserve the equilibrium geometry. Consequently, we recover the property that in a stable molecule some atoms like to accompany others, in other words, there are special relationships between certain atoms that the pioneers of chemistry called bonds. We are thus able to retrieve these 'special links' between atoms from the shape of the potential energy curve. Conversely, the energy curve of a system of two positive point charges for example does not correspond to an attractive energy profile but is *repulsive*, that is $d^2E(R)/dR^2 < 0$. As a result, when one charge is translated the other will *not* follow; on the contrary, the energy will keep decreasing as the intercharge distance increases.

Matters are simple for diatomics because there is only one possible bond. The energy curve of a bound diatomic automatically converted its attractive energy profile into a 'special relationship' between the two atoms present. Can we apply this principle to polyatomic molecules? The energy curve now becomes a *potential energy surface* in $3N-6$ dimensions (for non-linear molecules) for a molecule made up of N atoms. As before, we can find an equilibrium geometry signifying that the particular geometrical arrangement of the atoms has a lower energy than the free atoms at infinite separation. The key question is how can we express the stable, bound state in terms of 'special interactions or links' between pairs of atoms. Put differently, what gives us the right to call certain pairs of atoms bonded based on the shape of the total molecular energy surface? Clearly, what is needed is some systematic way of breaking down the multidimensional energy surface into one-dimensional energy curves, each curve corresponding to a bonded pair of atoms. We know of no elementary method that has been proposed or that could be presented at the level of this book. There is an alternative though. AIM is able to point out which pairs of atoms enjoy a 'special interaction'. Using the electron density, AIM can tell us precisely which atoms are bonded, and which are not, in any molecule. How this is done is explained in the next two sections.

4.3 The bond critical point

In section 2.4 we encountered a CP between the carbon and the oxygen nucleus in methanal (Figure 2.6). Although we left the meaning and proper name for this type of CP shrouded in mystery, many probably surmised that this CP should be associated with a bond. It will therefore not come as a surprise that this CP is called a *bond critical point* (BCP). When the complete set of all types of CP is discussed in Chapter 5, we will introduce an alternative notation for the BCP. As explained previously, a bundle of gradient paths originating at infinity terminates at the BCP and in that sense the BCP is the centre of the IAS. The BCP is not just a remarkable point of the IAS in a topological sense though: it is also at the BCP where ρ reaches its maximum within the IAS. In other words, any displacement away from the BCP but contained in the IAS reduces ρ from the value it possesses at the BCP, which we denote by ρ_{BCP} or simply ρ_b. Approaching infinitesimally close to the BCP along the IAS is equivalent to moving in a plane tangential to the IAS at the BCP. It is important to realize that the BCP is a *maximum* in the two dimensions forming this tangent plane. However, the BCP is *not* a maximum of ρ in the direction perpendicular to the tangential plane because then the BCP would be a maximum in three dimensions, and thus identical to the type coinciding with the nucleus. Instead, the BCP is a minimum of ρ in this orthogonal direction and can therefore be called a 3D *saddle point*. This will be elaborated in a more systematic and mathematical way in Chapter 5.

The BCP is a remarkable point topologically because it acts as a gateway between two bonded atoms. This can be vividly illustrated returning to our metaphor of someone travelling along gradient paths inside an atomic basin. Near the atom's IAS the traveller will graze the IAS, turn sharply in the immediate vicinity of the BCP and then be absorbed by the nucleus (see Figure 2.6 for example). When moving on the IAS itself, the traveller will be directed towards the BCP by the gradient vector until the BCP is reached. Upon arrival at the BCP the gradient vanishes and the traveller will not be forced to move in any particular direction. This is the only point where he could 'step into' the other atom. To push the metaphor, we can say that if a molecule is a house with rooms then the BCPs are the doors. We caution the reader not to think that electrons move only along gradient paths. Electrons move freely through IASs from one atom into another. Our metaphor applies only to the topology of the gradient vector field.

Bond critical points are also noteworthy in a *non*-topological sense. As explained in Chapter 10, numerous properties (scalar functions such as ρ and its derivatives) can be evaluated at the BCP and used to characterize the bond. There is ample evidence that the electronic structure of a molecule can be expressed accurately, reliably and compactly purely on the basis of BCP properties.

With an eye on ρ contour diagrams such as in Figure 1.1 or Figure 2.5, another interesting property of the BCP is disclosed. Outer contour lines, the ones of low electron density, typically encircle the whole molecule, whereas the contour lines of very high electron density enclose only one nucleus. Suppose we concentrate on a diatomic and draw ρ contour lines going from low to high ρ values, then at which ρ value do the contour lines split into two, one encircling each nucleus? This happens when ρ is equal to ρ_b, the electron density at the BCP. As a result, the BCP lies on a contour line that *just* encircles both nuclei. If the pattern of nested contour lines that encircle only *one* nucleus represents an 'atomic zone' (not to be confused with the AIM atom) then the BCP lies at the point where two 'atomic zones' touch. Loosely speaking, this observation points towards the statement that the BCP is the communication point or gateway between two bonded atoms.

4.4 The bond path

Let us focus on a particular BCP appearing in a polyatomic molecule. Can we conclude from the proximity of a BCP to the two nearest nuclei that the two atoms containing these nuclei are bonded? Fortunately there is direct topological evidence that enables one to conclude which atoms are bonded. As we now explain there is a line that literally links nuclei. Figure 4.2 represents a BCP appearing somewhere between the two nuclei *A* and *B*. It shows several gradient paths terminating at the BCP in a somewhat exaggerated plane tangential to the IAS as introduced in the previous section. This figure also illustrates that there is a gradient path *originating* at the BCP and terminating at nucleus *A*. Equally there is another gradient path also originating at the BCP but terminating at nucleus *B*. This set of two gradient paths is called an *atomic interaction line* (AIL). It is the rigorous topological expression that the two nuclei *A* and *B* are linked via a BCP. The AIL is found between every pair of nuclei whose atomic basins share a common IAS. Note that the presence of an AIL between two nuclei does *not* necessarily mean that they are bonded. As explained below we need one extra condition.

Let us first discuss a few properties of the AIL. The AIL is a line through the electron density along which ρ is a maximum with respect to any

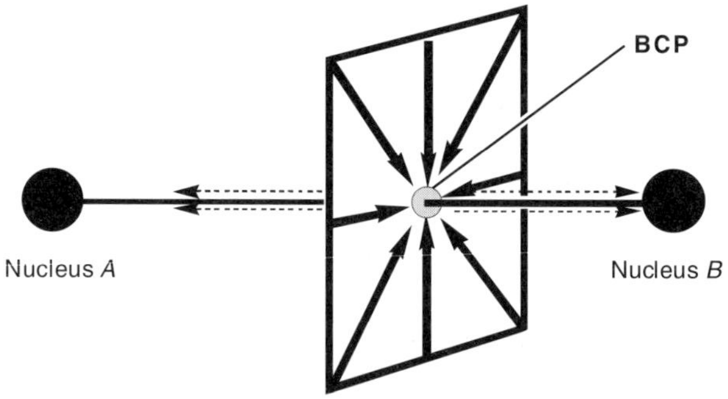

Figure 4.2 A schematic representation of a BCP in between two nuclei *A* and *B*. The curve linking *A* and *B* (the bond path) is not necessarily a straight line. The electron distribution ρ increases towards the BCP in a plane locally perpendicular to the bond path. Note that at the BCP ρ is a minimum along the bond path. (Adapted with permission from *J. Phys. Chem. A*, 1999, **103**, 2883–2890. Copyright 1999 American Chemical Society.)

neighbouring line. This fact is compatible with the situation at the BCP which lies on the AIL. It was mentioned in the previous section that ρ is a maximum at the BCP with respect to any lateral displacement in the plane tangential to the IAS. In that section it was also mentioned that for a motion in a direction orthogonal to the tangent plane, ρ is a minimum at the BCP. Since the AIL is locally perpendicular to the IAS at the BCP, ρ is a minimum at the BCP along the AIL. So, starting from the BCP the electron density increases approaching either nucleus. It is instructive to compare the electron density with a mathematically similar scalar field such as the temperature. The nuclei then match very hot points where the temperature is a maximum in any direction. These points are connected via a hot wire which corresponds to the AIL. The coolest point on this wire corresponds to the BCP and any lateral displacement away from the hot wire leads to a lower temperature.

As stated above, the existence of an electron density accumulation between nuclei along the length of the AIL is *not enough* to call them bonded. One external condition (i.e. not observable in ρ) that needs to be imposed on the nuclei is that the forces on them vanish. The reason for this extra condition becomes clear with an example. Consider a simple diatomic molecule consisting of two noble gas atoms. The attracting interaction responsible for the stability of such a diatomic is called dispersion. Without this interaction the two atoms would wander off to infinity and never end up in a finite equilibrium geometry, forming a stable molecule. Now we look at the theoretical description of such a diatomic. A computational scheme which ignores electron correlation does *not* predict a finite equilibrium geometry. Such a scheme is known to produce an electron density in which we observe an AIL, but we also know that the system is not bonded. Indeed, one discovers an AIL between the two noble gas atoms at *any* separation, but the atoms never form a molecule with a finite equilibrium geometry. As a result we cannot invoke the presence of an AIL as an indicator for bonding. What we need is an extra condition, based not on density but on energy considerations. This argument explains and justifies the reason and the nature of the extra condition we introduce above. We need one further refinement in order to make everything consistent. The second condition is that the whole molecule is in an energy minimum of its energy surface. This condition replaces the one of the vanishing forces because if a molecule is at an energy minimum all forces automatically vanish.

The full reason why this condition is imposed onto the AIL to conclude upon bonding is actually stronger than apparent from the example above. Indeed, AILs appear and disappear by changing the nuclear configuration of a molecule (see Chapter 6). In other words, by introducing large forces on the nuclei of a molecule one can almost create and destroy AILs *ad libitum*. At the time of writing, there is no theory that interprets AILs in terms of chemical bonding in molecular geometries seriously deviating from equilibrium. In the absence of such a theory we want to avoid a misinterpretation

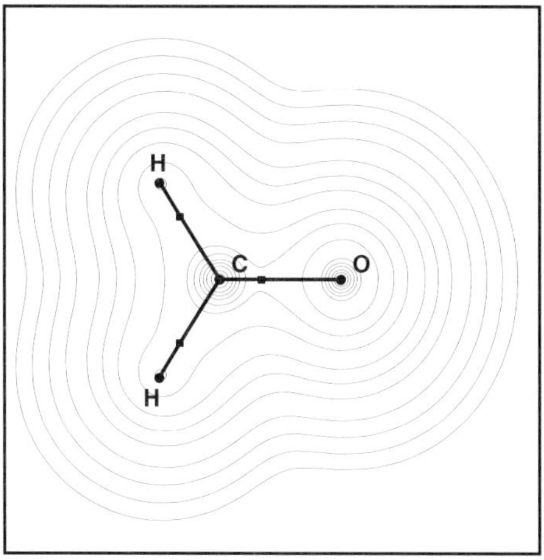

Figure 4.3 The molecular graph of methanal in its equilibrium geometry, superimposed on a contour map of ρ. The squares represent the BCPs appearing along the bond paths constituting the molecular graph.

of AILs in terms of bonding by imposing the condition that the molecule has to be at a local minimum of the potential energy surface. This condition also protects us against a false interpretation of AILs in promolecules discussed in section 1.7.

If the previously discussed condition is satisfied then the AIL is called a *bond path* (BP). Consequently all BPs are AILs but not all AILs are BPs. In other words, the set of BPs is a subset of the set of AILs. The set of bond paths in a molecule at its equilibrium geometry is called the *molecular graph* (MG). This graph expresses succinctly how quantum mechanics decides which nuclei are bonded. Figure 4.3 shows the molecular graph of our familiar methanal example. In this example we recover what chemists have been calling bonds for practically a century. However, note that the classically assigned double bond in methanal is represented by one AIL. The topology does not distinguish multiple bonds from single bonds; this issue will be deferred to Chapter 10.

However, we immediately emphasize two important points. The concept of an AIM bond should be taken in its widest sense – as an expression of *attractive stability* between two atoms, not just as a straightforward, classical Lewis-like covalent or polar bond. With reference to the energy considerations of section 4.2, a bond is a special energy relationship between two atoms. It says in almost colloquial terms that 'two atoms are happy sitting where they are being part of a stable molecule, at *equilibrium*'. As such, two bonded atoms contribute to the overall stability of the molecule at equilibrium. One should always keep in mind that the internuclear distance of two bonded atoms is in principle a *balance* between opposing forces. The dynamics of the bond formation – which are governed by repelling and attracting interactions – should not be confused with the actual bonding scheme. In the latter, all mutually opposing forces have settled and the only observation that counts is that the one-dimensional energy profile corresponding to the bond is attractive and therefore restores the equilibrium distance upon distortion.

The second point is that the external (i.e. *non*-electron density) condition imposed on the AIL to turn it into a BP constrains the AIM concept of bonding to equilibrium geometries only. That one can define a bond is an important achievement because, as shown in the examples, AIM proves to be a handy tool for extracting bonding schemes from the electron density. However, AIM does not judge what is a bond or not in a non-equilibrium situation. Clearly there is an urge for further research to generalize AIM's concept of a bond to include non-equilibrium geometries as they appear in molecular vibrations, transition states and perhaps partially or poorly optimized geometries.

It should be noted that the IAS, the BCP and the BP always appear together as a topological triplet. Practically, the BCP is sought first, starting from a good guess, and then the orientation of the plane tangential to the IAS is found and the direction perpendicular to it. The IAS is then traced, gradient path by gradient path, from the BCP outwards – in the opposite direction to the gradient vector. The two halves of the AIL are traced, each starting from the BCP and terminating at the nuclei – in the direction of the gradient vector.

4.5 Bonds in typical molecules

Figure 4.4 shows MGs of fifteen well known molecules. The MGs of some simple molecules such as methane, ethyne, hydrogen peroxide confirm their generally accepted bonding scheme. It should again be emphasized that when tracing the bond paths no *a priori* bonding picture is imposed on any molecule. Any algorithm generating topological information from the electron density operates in an unbiased way. The bonding schemes thus obtained are in that sense true observations – even literally if the algorithm acts on an experimental (X-ray) electron density.

The set of molecules shown in Figure 4.4 includes diborane, B_2H_6, which was once infamous because its bridge structure, suggested on the basis of spectroscopic evidence, could not be readily explained in terms of the Lewis model and standard hybridized orbitals. If diborane were regarded as two BH_3 units connected via a boron–boron σ bond then the Lewis formula would be seven two-electron bonds, or fourteen electrons. However, only twelve bonding (valence) electrons are available. Longuet-Higgins rationalized diborane's structure by connecting the borons via two bridging hydrogens each bonded to two boron atoms. This solution necessitates the introduction

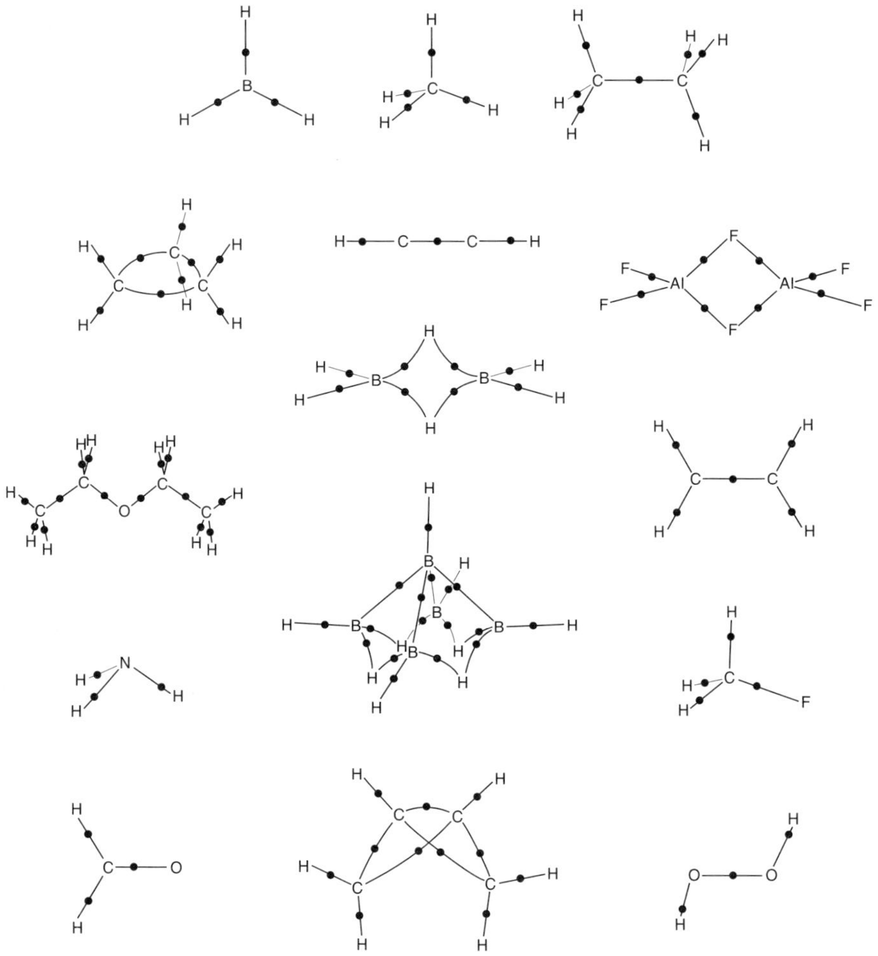

Figure 4.4 Molecular graphs for a set of molecules in their equilibrium geometries. A bond critical point is denoted by a black dot. (From Bader *et al.*, 1981. Reproduced by permission.)

of the new idea of a 'three-centre orbital' (encompassing the two boron nuclei and one bridging hydrogen) occupied by two electrons. Thus the classical [two-centre, two-electron] (2c-2e) counting rule which successfully elucidates structures with ordinary covalent bonds had to be modified to allow 3c-2e bonds. Similar considerations apply to molecules such as XeF_2, PF_5 or IF_7. Recently, researchers even proposed 5c-4e bonding in a rhodium-bismuth complex (Xu and Lin, 1998). Several other boranes (for example B_5H_9 in Figure 4.4) and carboboranes are all straightforward AIM applications but may cause problems within the valence bond picture. Of course, AIM does not yield any information about how many electrons should be

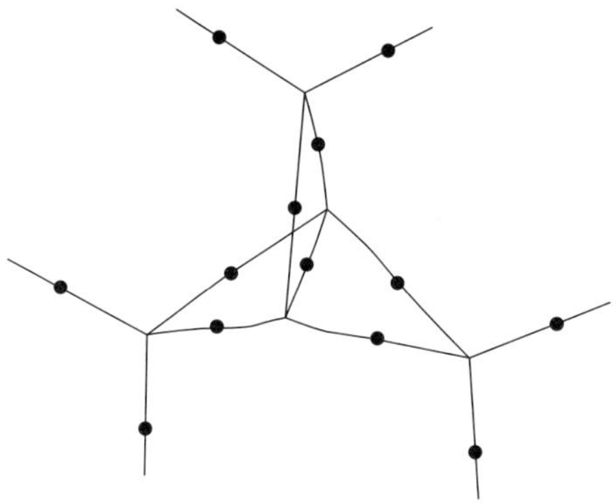

Figure 4.5 Planar projection of the molecular graph of [1.1.1]propellane generated from a theoretical electron density. The small black dots denote the BCPs. Note that the two bridgehead carbons have inverted geometries. (From Bader *et al.*, 1981. Reproduced by permission.)

assigned to a bond. Is this question relevant, however, other than in the classical context of determining which atoms are bonded? As stated before, the question about which atoms are bonded in a molecule can be answered unambiguously by AIM, given equilibrium geometries. In a next step, bonds established via an AIM analysis of ρ can be characterized by evaluating properties at the BCP (Chapter 10). In summary, AIM independently denies or confirms the bonding scheme suggested by the electron-centre counting rules.

A second important comment emanating from Figure 4.4 is that BPs are not always straight lines. Note, for example, the considerable curvature in the cyclopropane carbon–carbon bond paths. These curved BPs are consistent with the concept of bond strain. The deviation of BPs from straight lines can be quantified by measuring the bond path length R_b, the distance between two nuclei measured *along* the BP. The bond path length, R_b, is always greater than or equal to the bond length, R_e, which is simply the distance between the bonded nuclei, measured along a straight line.

A large number of saturated hydrocarbons have been subject to a topological analysis, including both branched and unbranched, cyclic and bicyclic hydrocarbons, and even highly strained systems such as tetrahedrane, cubane and spiropentane. The MGs of all these molecules are in complete agreement with the structures that have been assigned on the basis of chemical information and models of directed valency, augmented with the concept of bent bonds. However, existing models are inadequate for the so-called *propellanes*. An example of a propellane, called [1.1.1]propellane is shown in Figure 4.5. This hydrocarbon contains three three-membered rings all attached to two bridgehead carbon nuclei which are linked by a BP. This results in a structure with four BPs terminating at each of the bridgehead nuclei, which therefore have an 'inverted' geometry rather than a near-tetrahedral geometry expected of saturated carbons. The model of hybridized orbitals cannot describe such a situation, even assuming bent bonds. The use of AIM is further illustrated for molecules obtained via a systematic addition of one carbon to each three-membered ring in [1.1.1]propellane. As such, we obtain compounds called [2.1.1]propellane, [2.2.1]propellane and [2.2.2]propellane, which all possess a bond between their bridgehead carbons.

4.6 Bonds in van der Waals molecules

Van der Waals forces are the interactions between closed-shell molecules, consisting of repulsive and attractive contributions. The former arise from coulombic repulsions and, indirectly from the Pauli principle via charge-cloud overlap effects. The latter arise from the three possible combinations of interactions between permanent and induced multipoles, i.e. multipole–multipole (for example, dipole–dipole), multipole–induced multipole and induced multipole–induced multipole. The latter two contributions involve so-called polarizabilities. The van der Waals forces are responsible for the stability of complexes of atoms and/or molecules such as Ar_2, $Ar–SO_2$, $H_2O–H_2O$. These complexes can be studied experimentally using sophisticated spectroscopic techniques and 'molecular beams'. Hydrogen-bond complexes are also governed by van der Waals forces, but in view of their ubiquity and general importance they will be discussed separately in Chapter 10.

Figure 4.6 shows the CO_2 dimer as an example of a van der Waals molecule. The nuclear framework corresponds to the lowest energy or global minimum of this complex. Based on the energy considerations of section 4.2, we infer that at least one pair of nuclei must be linked by a bond to express the attractive attraction between the two carbon monoxide monomers. It turns out that there is one such bond, between the oxygens of different carbon dioxide molecules. This is a direct illustration that BPs cover a more general concept of bonding than that captured by the Lewis-like picture, in which the bond is represented by a line.

Simple dimers such as Ar_2 clarify the necessity to impose onto the AIL the conditions of vanishing forces and of the energy's positive second derivative, in order for it to become a BP. The point is that the presence of an AIL does *not* imply a bond whereas a BP does. Let us consider the argon dimer Ar_2 which is held together entirely by *dispersion*, which is a contribution to the total van der Waals energy of the type induced multipole–induced multipole. Indeed, some calculations performed at inadequate levels of theory fail to recover the bound state of Ar_2 because they do not reproduce the dispersion energy correctly. However, the wave function obtained at such an inadequate level yields an AIL between the argon nuclei. Clearly this AIL cannot be interpreted as a bond because the argon atoms wish to move

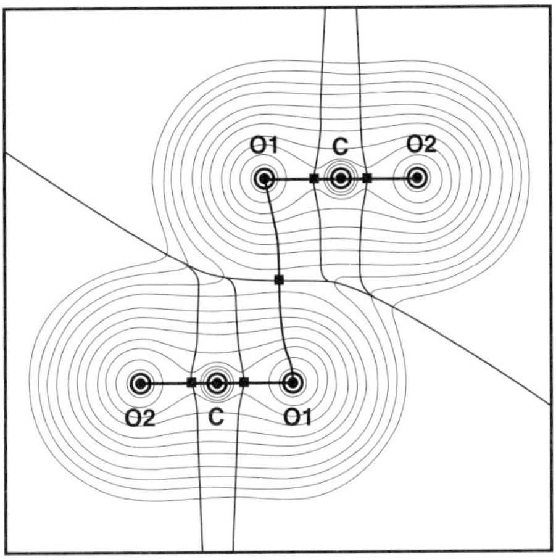

Figure 4.6 Superposition of an electron density contour map (thin) and the molecular graph (bold) of the global minimum (C_{2h} symmetry) of the CO_2 dimer. (From Bone and Bader, 1996. Copyright © American Chemical Society; reproduced by permission.)

infinitely far apart. This is why the two extra conditions (which can be taken as one), drawn from the energy profile, are added to the topological condition of the presence of an AIL which is drawn from the electron density.

4.7 Bonds in organolithium molecules

Organolithium compounds have unusual geometries that challenge traditional ideas of bonding and are therefore an ideal testing ground for AIM. Our theory confirms the recognized fact – obtained by non-AIM considerations – that the carbon–lithium bond is predominately ionic (details in section 10.2). Furthermore, AIM provides the bonding schemes for a considerable survey of organolithium molecules as straightforward applications.

The definition of *coordination number*, which is important in inorganic chemistry, surfaces as a trivial corollary of the definition of a bond from the presence of a BP. Indeed, according to AIM the coordination number of an atom is simply taken to be the number of BPs terminating at the nucleus that it contains. Interestingly, many hypercoordinated carbons have been detected (penta, hexa, hepta, . . .) (Olah and Rasul, 1997).

Figure 4.7 shows an example of the organolithium compound $(LiCH_3)_2$ which confirms two observations that have been made in a survey of a few dozen organolithium molecules (Ritchie and Bachrach, 1987). The occurrence of very long carbon–carbon bonds (just under 7 Bohr) is probably best described as of van der Waals type in view of the large diffuse charge

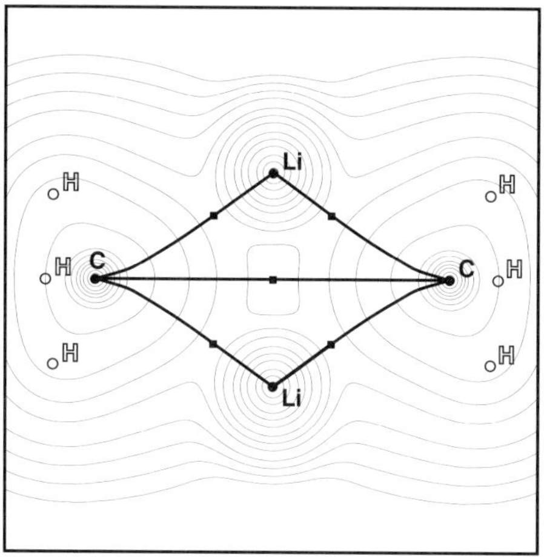

Figure 4.7 Superposition of an electron density contour map and the MG of an equilibrium geometry of the $LiCH_3$ dimer. This molecule can be viewed as ethane severely perturbed by two lithium atoms.

that lithium induces in the methyl groups (and other molecular fragments). The second important deduction is that a Li–Li bond does *not* exist, i.e. a BP linking two lithium nuclei has never been encountered. Surely one can argue that this is inconsistent with the existence of a bound state of the lithium dimer, Li_2, discussed in section 2.6. But even here the rule that lithium atoms never directly bind is obeyed because each lithium atom is bonded to the NNA in the middle. In other words, both lithium atoms lose some electronic charge (since, they are very electropositive) and 'create' an object in the middle which looks like a charge nodule to which they both bind. Note that this whole discussion of the lithium dimer holds because this topology occurs at its *equilibrium* geometry.

4.8 The agostic bond

As a final example we discuss the *agostic bond* emerging from work in the area of organometallic chemistry (Brookhart and Green, 1983). It typically occurs between a hydrogen and a transition metal where the hydrogen is bonded to a carbon and is therefore denoted by C–H · · · M. The first anomalies in experimental data suggesting the existence of a special interaction between a metal and a hydrocarbon ligand date from the late 1960s. Further confirmation from NMR and IR spectra as well as X-ray and neutron diffraction experiments kept accumulating until the bond was given its current name in the early 1980s.

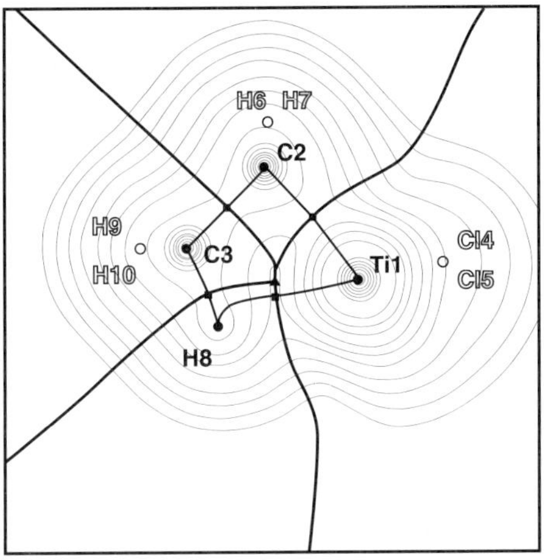

Figure 4.8 Superposition of an electron density contour map, the MG and IASs in the symmetry plane of $C_2H_5TiCl_2^+$. The BCPs are denoted by squares. The labels of the nuclei that lie in the mirror plane are bold and those that do not lie in this plane are open. The bond path corresponding to the agostic bond is quite curved near the agostic hydrogen H8. (From Popelier and Logothetis, 1998. Reproduced by permission.)

Figure 4.8 shows the independent topological evidence that AIM presents for an agostic bond in $C_2H_5TiCl_2^+$ embodied in the BP linking Ti and H8. It is believed that in this system the agostic bond is important in understanding the catalysis of Ziegler–Natta polymerization. A similar agostic bond was found in $C_3H_7TiCl_2^+$, which has a hydrocarbon ligand longer by one methylene group. Depending on the conformation of the terminal methyl group the Ti is either bonded to the hydrogen or to the carbon, which then rigorously becomes five-coordinated. In fact, the rapid spinning of this methyl group around its C–C axis entails a change in topological connectivity (see Chapter 6). The dynamic character of the interaction between Ti and C or H is probably why many experts seem to think of an agostic bond as a σ interaction with the entire C–H group.

In closing this chapter one important general issue must be mentioned because it arose in the study of agostic bonds and has appeared in other systems such as organolithiums. There are consistent indications that some interactions between atoms are definitely present but are topologically *not* expressed. In other words, one can infer from anomalies of the molecule's geometry that two atoms are interacting or strongly influencing each other but one does not observe a bond path between them. It is hoped that a full understanding of this phenomenon will be forthcoming from the careful study of ρ's topology for non-equilibrium geometries.

Summary

- The concept of the chemical bond originates from valence considerations and has been refined using simple orbital pictures and knowledge of molecular geometry. The mainstream idea of a bond is currently *not* based on solutions of the Schrödinger equation.
- Atoms in molecules (AIM) precisely and rigorously defines a bond fully compatible with both calculated wave functions and experimental densities.
- An atomic interaction line (AIL) points out a special interaction or link between two nuclei and is a line of maximum electron density with respect to neighbouring lines.
- Two atoms are bonded if their nuclei are linked via a bond path (BP). A BP is an AIL in a molecule in a (stable) minimum energy conformation. A collection of BPs is called a molecular graph (MG).
- The bond critical point (BCP) is the 'communication point' between two atoms, and bonds can be characterized via properties evaluated at the BCP.
- MGs recover the conventional picture of bonding in many well known molecules but arbitrate in a straightforward manner in cases where conventional methods break down.
- The coordination number of an atom is defined as the number of BPs terminating at the nucleus it contains.
- Bonding can be conveniently studied using AIM in any chemical discipline as long as a wave function or experimental density can be obtained. Examples discussed include van der Waals bonds, agostic bonds, hydrogen bonds and organolithiums and borane molecules.

References

Bader R. F. W., Nguyen-Dang T. and Tal Y. (1981). *Rep. Prog. Phys.*, **44**, 893–948.

Bone R. G. A. and Bader R. F. W. (1996). *Journal of Physical Chemistry*, **100**, 10892–10911.

Brookhart M. and Green M. L. H. (1983). Carbon–hydrogen–transition metal bonds. *Journal of Organometallic Chemistry*, **250**, 395–408.

Olah G. A. and Rasul G. (1997). From Kekulé's tetravalent methane to five-, six-, and seven-coordinate protonated methanes. *Accounts of Chemical Research*, **30**, 245–250.

Popelier P. L. A. (1999) Quantum molecular similarity. 1. BCP space. *Journal of Physical Chemistry A*, **103**, 2883–2890.

Popelier P. L. A. and Logothetis G. (1998). Characterization of an agostic bond on the basis of the electron density. *Journal of Organometallic Chemistry*, **555**, 101–111.

Ritchie J. P. and Bachrach S. M. (1987). Bond paths and bond properties of carbon–lithium bonds. *Journal of the American Chemical Society*, **109**, 5909–5916.

Xu Z. and Lin Z. (1998). Unusual five-center, four-electron bonding in rhodium–bismuth complex with pentagonal-bipyramidal geometry. *Angewandte Chemie* (International edition), **37**, 1686–1689.

Further reading

Murrell J. N., Kettle S. F. A. and Tedder J. M. (1985). *The Chemical Bond*. Chichester: Wiley.

Runtz G. G., Bader R. F. W. and Messer R. R. (1977). Definition of bond paths and bond directions in terms of the molecular charge distribution. *Canadian Journal of Chemistry*, **55**, 3040–3045.

Webster B. (1990). *Chemical Bonding Theory*. Oxford: Blackwell Scientific.

Chapter 5

The full topology

What else could involve rings and cages?
A circus

5.1 The ring critical point

So far we have encountered two types of critical points: the BCP and the maximum, which either coincides with a nucleus or exists on its own, without a nucleus. What happens if nuclei are topologically connected to form a ring structure such as in benzene? Such a connectivity scheme gives rise to a new type of CP called a *ring critical point* (RCP). A precise mathematical definition of this new type of CP is given in section 5.3 but here we describe it by means of an example.

Figure 5.1 shows two cross-sections of the gradient vector field of benzene, one in the molecule's symmetry plane and one perpendicular to this plane. As suggested by Figure 5.1(a) an infinite number of gradient paths originate at the RCP and terminate at a nucleus or at a BCP. This set of gradient paths defines a surface called the *ring surface*. Because of symmetry constraints, the ring surface in benzene is flat but this is not necessarily so. Since gradient paths always move from a low electron density to a higher one the RCP must be a point of minimum electron density within the ring surface. A *ring* is defined as a part of a molecular graph (MG) which bounds a ring surface. In the case of benzene the ring consists of the set of six C–C bond paths (BPs).

It should be emphasized that a ring is topologically as meaningful as a BP, i.e. if AIM predicts BPs where we expect them by conventional arguments then the expected concomitant ring will be recovered as well, as in the case of benzene. However, for molecules stimulating debate such as organolithiums the detection of a (topological) ring may come as a genuine prediction. Figure 5.1(b) highlights a complementary aspect of the topology around the RCP. There are two gradient paths originating at infinity and terminating at the RCP. They constitute what is called the *ring line*. In the case of benzene, the ring line is a straight line perpendicular to the ring surface. It should be also noted that the RCP lies in the centre of the benzene ring owing to symmetry constraints. An example of an RCP lying in

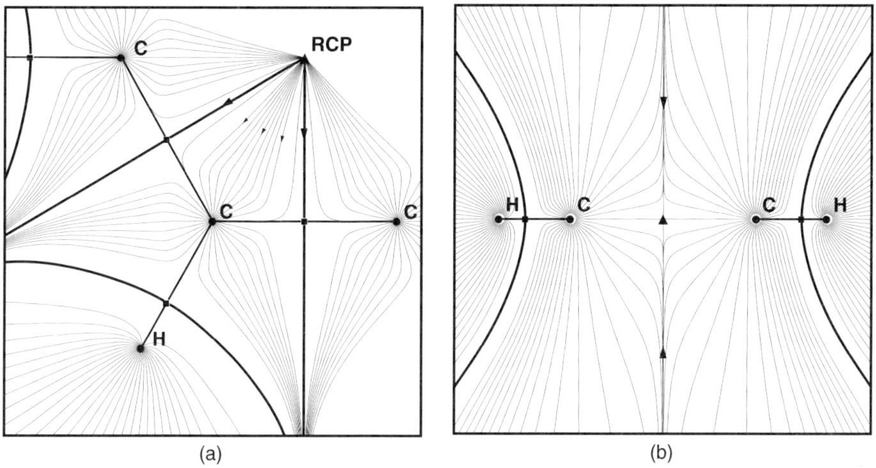

Figure 5.1 (a) A portion of the gradient vector field in the symmetry plane of benzene. A relevant part of the molecular graph (MG) and relevant interatomic surfaces (IASs) are shown. A set of gradient paths attracted to some carbon nuclei and one hydrogen nucleus is also shown. The BCPs are marked by a black dot and the RCP by a small triangle. The arrows mark the orientation of the gradient paths. Note that an infinite number of gradient paths originate at the RCP. (b) A representation of the gradient vector field in a plane through the RCP and perpendicular to the symmetry plane of benzene. The ring line appears in the middle of the plot as a vertical line going through the RCP. The ring line consists of two gradient paths, each of which originates at infinity and terminates at the RCP.

the centre of a four-membered ring involving the two borons and the two bridging hydrogen atoms is found in diborane. In the absence of symmetry, an RCP can be found anywhere inside the ring, for example, in a five-membered ring occurring in the so-called C_5 conformation of an amino acid in a peptide (see Figure 10.7a in Chapter 10). Of course, the smallest ring requires three atoms, such as in cyclopropane, but there is no upper limit in the number of nuclei participating in a ring. Rings containing as many as thirteen atoms have been analysed topologically (see Figure 10.7b).

It is instructive to point out that an RCP is in many respects the inverse of the BCP because the two appear as complementary types of saddle points. Recall that the BCP is a maximum in the interatomic surface (IAS) and a minimum with respect to the bond path. The RCP, on the other hand, is a minimum in the ring surface and a maximum on the ring line. If we compare the IAS to the ring surface (because they are both two-dimensional topological manifolds) then the BCP is a *maximum* with respect to the IAS and the RCP is a *minimum* with respect to the ring surface. Similarly, if the BP is equated to the ring line, then the BCP is a *minimum* with respect to the BP and the RCP is a *maximum* with respect to the ring line.

5.2 The cage critical point

The final type of CP is a minimum in three dimensions and is therefore the perfect counterpart of the (N)NA critical point. It is called a *cage critical point* (CCP) and appears, for example, in the strained hydrocarbon tetrahedrane, C_4H_4, which is shown in Figure 5.2 (see Plate section). Clearly the BPs mark the boundaries of the four ring surfaces that enclose the interior of the molecule as a cage, hence the name cage critical point. No gradient paths can terminate at a CCP, they can only originate there and be attracted to a nucleus, a BCP or a RCP, as illustrated in Figure 5.2. We define a *cage* as that part of an MG which contains at least two rings, such that the union of the ring surfaces bounds a region in real 3D space which contains a CCP. While it is mathematically possible for a cage to be bounded by only two ring surfaces, the minimum number found in an actual molecule so far is three, as in bicyclo[1.1.1]pentane. This molecule is the same as the previously discussed molecule [1.1.1]propellane (see Figure 4.5) but has hydrogens attached to the bridgehead carbons thereby breaking the bridgehead carbon–carbon bond.

5.3 Review of all types of critical point

We are now in a position to put the four types of CP into a common framework and to describe them using the mathematics presented in Box 5.1. The CPs are classified according to their *spectrum*, which is the set of eigenvalues of the Hessian matrix $\nabla\nabla\rho$. The number of non-zero eigenvalues and their associated signs are sufficient to define the type of a CP. The *rank* of a CP, denoted by r, is the number of non-zero eigenvalues of the spectrum. Recall that a vanishing eigenvalue or zero curvature means that a function is 'locally flat', i.e. it is neither a minimum nor a maximum in the direction of the corresponding eigenvector. In that direction (or dimension), a point where the second derivative vanishes is called a *point of inflexion*. The *signature* of a CP, denoted by s, is a number that keeps track of three eigenvalue signs of the spectrum. The number -1 is assigned to a negative eigenvalue and $+1$ to a positive one. The signature is then defined as the sum of these three numbers. For example, at the BCP the Hessian possesses one positive and two negative eigenvalues and the signature equals $(+1)+(-1)+(-1)=-1$. The doublet of values that characterizes a CP, being the rank and the signature, is conveniently expressed as (r, s).

Table 5.1 summarizes the four types of CPs in terms of their topological acronym (BCP, RCP, etc.), spectrum$\{\lambda_1, \lambda_2, \lambda_3\}$and (r, s) notation. It is hard to visualize all three curvatures of a 3D function in the plane of the paper, but Figure 5.3 attempts to give an impression of a 3D maximum, minimum and saddle point. In this figure the scatter dot density suggests whether the CP is a maximum or a minimum with respect to a plane spanned by two eigenvectors. For convenience such a plane is called an *eigenplane*. The

Box 5.1 A closer look

Eigenvalues and eigenvectors of the Hessian

The mathematical concepts of eigenvalues and eigenvectors appear in many areas of physics and chemistry but rather than presenting a general discussion we briefly review them directly in the context of the electron density. The *Hessian* is a matrix of second derivatives of some function with respect to all possible combinations of coordinates. For our purposes we consider a 3×3 Hessian of the electron density, denoted by $\nabla\nabla\rho$ and expressed explicitly as:

$$\nabla\nabla\rho = \begin{Bmatrix} \dfrac{\partial^2\rho}{\partial x^2} & \dfrac{\partial^2\rho}{\partial x\partial y} & \dfrac{\partial^2\rho}{\partial x\partial z} \\[2ex] \dfrac{\partial^2\rho}{\partial y\partial x} & \dfrac{\partial^2\rho}{\partial y^2} & \dfrac{\partial^2\rho}{\partial y\partial z} \\[2ex] \dfrac{\partial^2\rho}{\partial z\partial x} & \dfrac{\partial^2\rho}{\partial z\partial y} & \dfrac{\partial^2\rho}{\partial z^2} \end{Bmatrix}$$

This matrix can be evaluated at any point in space (provided the derivatives are defined) but we focus on $\nabla\nabla\rho$ being evaluated at a BCP. Each matrix element should be interpreted as a curvature despite its being hard to do so visually. Also the actual matrix values depend on the choice of the coordinate system (x, y, z). The *intrinsic* structure or profile of ρ does not depend on the choice of coordinate system because no matter how a molecule is rotated (or translated) its wave function, and therefore electron density, remains unaltered. To eliminate the arbitrariness of the Hessian's representation we find the eigenvalues and corresponding eigenvectors of $\nabla\nabla\rho$ at a given BCP. It is a well known property that eigenvalues are invariant with respect to rotation and are thus ideal to reveal the internal structure of the Hessian. The eigenvalue equation has three solutions, each of which obeys the following matrix equation:

$$(\nabla\nabla\rho)\ \mathbf{u}_i = \lambda_i \mathbf{u}_i \quad (i = 1, 2, 3)$$

where λ_i is the ith eigenvalue and $\mathbf{u}_i$ is a column vector with three components representing the associated eigenvector.

The three eigenvectors are mutually orthogonal and coincide with the so-called *principal axes of curvature*. This name refers to the fact that the magnitudes of the three second derivatives of ρ calculated with respect to these axes are extremized. Moreover, the three eigenvalues are always real and are the only non-vanishing curvatures, compared with the nine Hessian matrix elements found before. In other words, the Hessian becomes a diagonal matrix when expressed with respect to the axes system constituted by the principal axes of curvature. Practically speaking, then, each eigenvector represents an axis and the corresponding eigenvalue determines the profile of the electron density along this axis. For example, if one of the eigenvectors at the BCP is locally directed along the BP and if its associated eigenvalue is positive then we know that ρ is a minimum along the BP's direction (this follows simply from the interpretation of second derivatives).

It is instructive to introduce an important function which will be discussed extensively in later chapters. This is the *Laplacian of the electron density* and is denoted by $\nabla^2\rho$. It can be defined as the sum of the eigenvalues of the Hessian, or $\nabla^2\rho = \lambda_1 + \lambda_2 + \lambda_3$. The Laplacian of ρ is invariant with respect to coordinate transformation and is therefore also an intrinsic property of the electron density.

Table 5.1 Summary of all four types of CP in terms of their full name, acronym, eigenvalue spectrum and (*r*, *s*) notation.

Name	Acronym	λ_1	λ_2	λ_3	(*r*, *s*)
Cage critical point	CCP	+	+	+	(3, +3)
Ring critical point	RCP	–	+	+	(3, +1)
Bond critical point	BCP	–	–	+	(3, –1)
(Non) nuclear attractor	(N)NA	–	–	–	(3, –3)

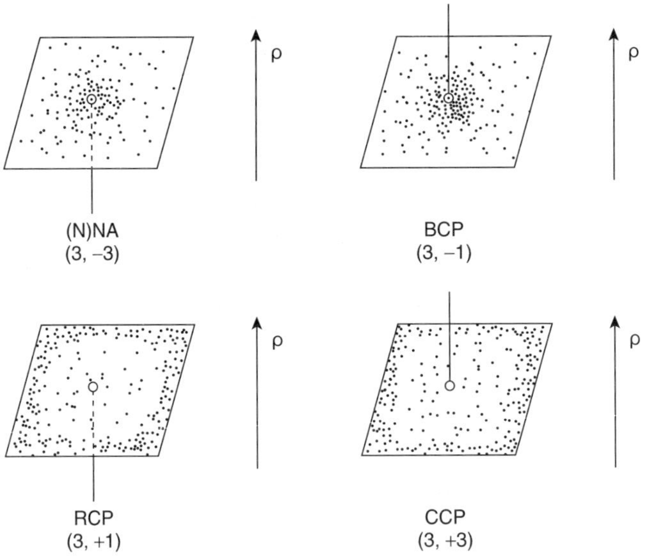

Figure 5.3 Compound representation of the four types of CP in the electron density. Each picture depicts the spectrum of the Hessian at the CP by showing in which direction(s) the CP is a minimum or a maximum. The pictures are somewhat distorted in that the vertical dimension is also used to mark the value of ρ. The density of scatter dots in the eigenplane implies whether the CP is a minimum or maximum in that plane. If the line perpendicular to the eigenplane is above the CP, then it is a minimum with respect to that direction and vice versa.

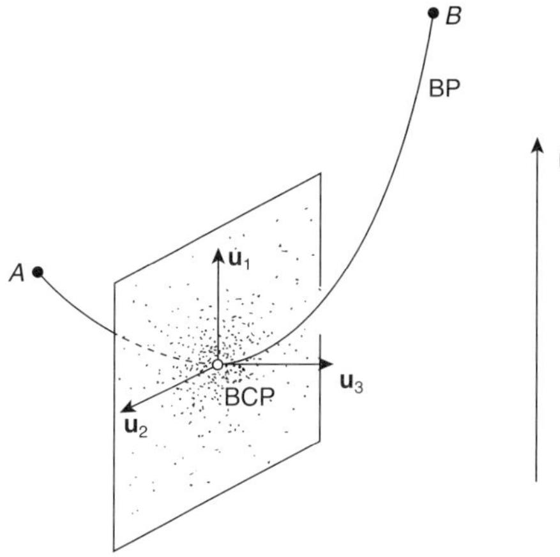

Figure 5.4 Representation of the local structure of the Hessian at the BCP. The two eigenvectors $\mathbf{u}_1$ and $\mathbf{u}_2$ span a plane in which the BCP is a maximum as marked by the density of scatter dots. This plane is called an eigenplane and is tangential to the interatomic surface (IAS). The remaining eigenvector $\mathbf{u}_3$ is tangential to the bond path (BP) linking nuclei *A* and *B*. The BCP is a minimum in this direction as shown by its position with respect to the 'ρ' axis.

behaviour of the function in the remaining direction orthogonal to the eigenplane is symbolically marked by a short line which lies above the eigenplane if the CP is a minimum and below it if the CP is a maximum.

Throughout the book we adopt the following convention to order the eigenvalues: $\lambda_1 \le \lambda_2 \le \lambda_3$. For example, the Hessian evaluated at the BCP has two negative eigenvalues, or $\lambda_1 \le \lambda_2 < 0$, and one positive eigenvalue, or $\lambda_3 > 0$. These two mathematical expressions are equivalent to the statements made previously about the saddle point character of the BCP. Indeed, looking at Figure 5.4 it becomes clear that the BCP is a maximum in the eigenplane spanned by the eigenvectors $\mathbf{u}_1$ and $\mathbf{u}_2$ and that it is a minimum with respect to the direction of the eigenvector $\mathbf{u}_3$. At the BCP, this eigenplane is tangential to the IAS and the eigenvector $\mathbf{u}_3$ is tangential to the BP.

Most molecules at or in the neighbourhood of energetically stable geometrical configurations of the nuclei are of rank three. A CP with rank strictly smaller than three ($r < 3$) is called *degenerate*. Such a CP has at least one zero curvature and is topologically unstable. This means that a small change in ρ, brought about by a displacement of the nuclei, causes the degenerate CP to vanish or to bifurcate into a number of non-degenerate or stable rank-three CPs. The details of such structural changes are given in the next chapter, but for the sake of completeness we briefly discuss all possible types of degenerate CP. A (degenerate) CP of rank 2 is much easier to imagine than one of rank 3 because we need only three dimensions instead of four. Based on the combinations shown in Table 5.1, it is straightforward to work out that there are only three types of rank-two CP: (2, −2), (2, 0) and (2, +2). No special names have been proposed for them, but it is clear that the first

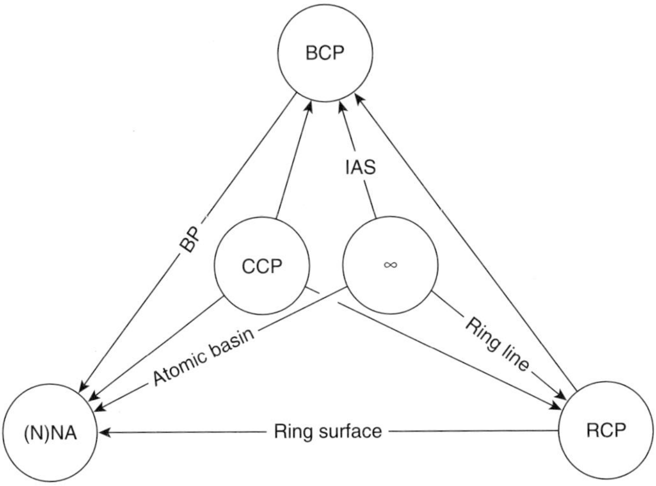

Figure 5.5 A schematic representation of the nine possible types of gradient path connecting two *different* types of CP. The diagram involves five possible points: (N)NA, BCP, RCP, CCP and infinity (∞). Each arrow points towards the terminus of a gradient path. Some gradient paths are part of a topological object with a specific name such as IAS, BCP, atomic basin, ring line and ring surface.

type is a maximum, the second is a simple 2D saddle point and the last one is a minimum. Similarly, the only rank-one CPs are denoted by (1, –1) (or maximum) and (1, +1) (or minimum). In one dimension, a saddle point situation is impossible. Finally, we should mention the (0, 0) CP in which all eigenvalues vanish, which means that the point is part of a ρ profile which is locally flat in *any* direction. Although such a highly degenerate CP may seem merely of academic interest it is worth discussing in the context of the following observation: *every gradient path originates at a CP and terminates at a CP.*

For a free molecule, ρ and all its derivatives, such as the gradient and the Hessian, vanish at infinity. Consequently there is a 'sea' of CPs at infinity, all of rank zero because all three eigenvalues vanish (the eigenvectors, however, are undetermined, i.e. they can point in any direction). At infinity, an infinite number of gradient paths originate, most being attracted to a nucleus and therefore terminating at a CP of type (3, –3). Another set of gradient paths originate at infinity but terminate at a BCP or (3, –1), yet others terminate at a RCP or (3, +1). The reader is reminded that a gradient path can never terminate at a cage critical point (CCP), which shares this property with any CP at infinity. Further possibilities encompass a gradient path going from a CCP to a RCP, BCP or (N)NA. A scheme of nine types of gradient path linking all possible types of CP (excluding those of ranks 1 and 2, and excluding gradient paths linking CPs of the same type) is shown in Figure 5.5. The fact that some gradient paths have not been given a special name is probably owing to their rarity.

5.4 The Poincaré–Hopf rule

There is a remarkable equation discovered many decades ago in the context of the mathematical branch of topology. It is a most general equation valid for any scalar function in three dimensions but we will discuss it here solely in the context of ρ. In our particular application the *Poincaré–Hopf relationship* links the number of non-degenerate CPs for a non-periodic systems as follows (thus excluding a crystal but including a free molecule or molecular complex):

$$n - b + r - c = 1$$

where n is the number of (N)NAs, b the number of BCPs, r of RCPs and c of CCPs. The collection of numbers *(n, b, r, c)* is called the *characteristic set*. Let us apply this equation in some simple examples. Once again we revisit methanal which has four nuclei ($n = 4$) and three BCPs ($b = 3$), one between C and O and one between C and each H. It is obvious that the Poincaré–Hopf relationship is satisfied because $4 - 3 = 1$. The characteristic set of benzene is (12, 12, 1, 0) which again satisfies the equation as $12 - 12 + 1 - 0 = 1$. The Li_2 molecule has two nuclei and one NNA yielding $n = 3$ and $b = 2$ such that $3 - 2 = 1$, again confirming the topological consistency of the molecular graph. Finally, the more complex topology of tetrahedrane once more obeys the Poincaré–Hopf relationship since $n - b + r - c = 8 - 10 + 4 - 1 = 1$.

One practical application of the Poincaré–Hopf rule is to check a calculated pattern of CPs for topological consistency. Indeed, locating CPs is not always that straightforward and a violation of the Poincaré–Hopf rule typically spurs a further hunt for missing CPs. However, a topological pattern that obeys this rule is not necessarily complete because it could happen that *both* a RCP and a BCP have not been detected. This can be understood by looking at the alternating sum appearing in the Poincaré–Hopf rule. If $n - b + r - c = 1$ holds, which corresponds to a complete and correct topology, then

$$n - (b - 1) + (r - 1) - c = n - b + r - c + (1 - 1) = 1$$

clearly holds as well. In summary, the Poincaré–Hopf rule can never prove the completeness of a set of CPs, it can only point out that it is inconsistent and therefore incomplete.

5.5 Ellipticity

Critical points can be characterized in a quantitative way by evaluating properties at them. It was mentioned in section 4.3 that the derivatives of the electron density can be used to characterize the BCP and thus the associated bond (to be discussed in detail in Chapter 10). The suggestion of utilizing the derivatives of ρ can be made more precise if we propose to use the

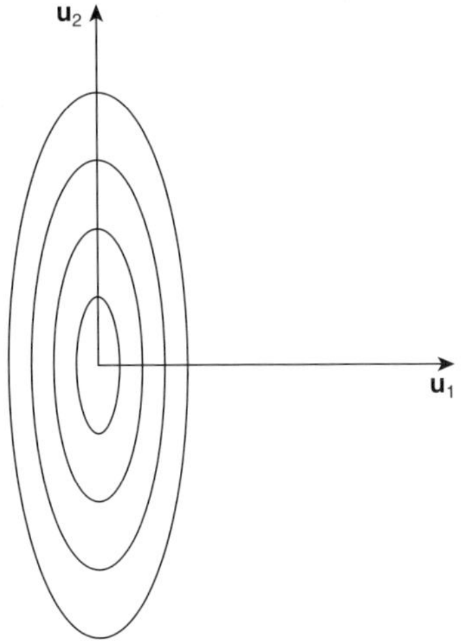

Figure 5.6 A contour diagram of the electron density in the eigenplane at a BCP spanned by the eigenvectors $\mathbf{u}_1$ (minor axis) and $\mathbf{u}_2$ (major axis). The number of contour lines traversed per unit length is larger in the direction of $\mathbf{u}_1$. Hence the curvature of ρ is larger in this direction which is compatible with λ_1 being more negative.

eigenvalues λ_1, λ_2 and λ_3 of the Hessian of ρ. Evidently we can employ the 'raw data' (λ_1, λ_2, λ_3) but it will prove to be a helpful complement to propose quantities *derived* from the three eigenvalues. Any quantity derived from the Hessian's eigenvalues, such as the Laplacian, adopts the property of being independent of the coordinate system used. Note that the new quantity we describe in the next paragraph applies only to BCPs.

In the eigenplane spanned by the two eigenvectors $\mathbf{u}_1$ and $\mathbf{u}_2$, two negative eigenvalues (λ_1 and λ_2) summarize the curvature of the electron density near the BCP. Since, by convention, $\lambda_1 \leq \lambda_2 < 0$, the absolute value of λ_1 is greater than that of λ_2, or $|\lambda_1| \geq |\lambda_2|$. As a result, the actual curvature in the direction of eigenvector $\mathbf{u}_1$ is larger in *absolute* terms. In other words, referring to the classical analogy with walking up a hill, the electron density 'hill' at the BCP is steeper in the direction of $\mathbf{u}_1$. Figure 5.6 prompts an equivalent interpretation by showing the number of ρ contours that are crossed per unit length in both eigenvector directions. Another useful term is that λ_1 is the 'hard' curvature whereas λ_2 is the 'soft' one. The ratio of $|\lambda_1|$ and $|\lambda_2|$ will then measure the disparity between the two orthogonal curvatures. The quotient $|\lambda_1|/|\lambda_2|$ in effect gauges how pronounced the elliptical shape of the contour diagram is, shown in Figure 5.6, and leads to the so-called *ellipticity* ε. This quantity is defined as $(\lambda_1/\lambda_2) - 1$ and ranges between zero and infinity.

The first term is always positive (because $\lambda_1 \leq \lambda_2 < 0$) and is manifestly larger than or equal to 1 because $|\lambda_1| \geq |\lambda_2|$. In order to retrieve a zero ellipticity for a circular contour diagram, the number 1 has been subtracted from the quotient. So cylindrical bonds, occurring in linear molecules, all have vanishing ellipticity. Figure 5.6 also illustrates that if the soft curvature becomes zero, $\lambda_2 \rightarrow 0$, the ellipse becomes infinitely 'thin' or 'pronounced' and accordingly $\varepsilon \rightarrow \infty$. In other words, the major axis of the ellipse becomes infinitely long and reveals that ρ does not change in this direction. The electron density has become flat in the direction of eigenvector $\mathbf{u}_2$. We postpone the practical applications of the ellipticity to later chapters. In particular, in Chapter 6, we will demonstrate the role the ellipticity plays in structural changes (CP annihilation). In Chapter 10, ε will be shown to relate to the π-character of a bond, in as far the π-character is reflected in the shape of ρ.

In this chapter we have expounded the full classification of CPs using the second derivatives of ρ. This is a further example of how 'introspection into the electron density via differentiation' helps us to extract information from ρ. This information will lead to a novel understanding of the properties of molecules, as shown in Chapter 8. Once more, we emphasize that at no point do we invoke an arbitrary reference density. All the information is lurking within the electron density itself; it can be recovered by applying some fairly elementary mathematics. The mathematical core of the classification and topological characterization of CPs can be transferred to scalar functions other than the electron density, such as the Laplacian (Chapter 8). The knowledge acquired in this chapter puts us in a perfect position to explore how a molecule's topology can *change*. This is the main theme of the next chapter.

Summary

- Critical points are classified using the Hessian of the electron density. This second derivative matrix yields three eigenvalues and three associated eigenvectors. The former express the local curvatures at the CP, the latter the corresponding directions of curvature.
- The rank of a CP is the number of non-zero eigenvalues and the signature the sum of their signs. Degenerate CPs have a rank lower than 3.
- The number of types of CP exceeds the rank of the CP by 1. As a result, there are four types of rank-three CP: (non)nuclear attractor (N)NA, bond critical point (BCP), ring critical point (RCP) and cage critical point (CCP).
- A degenerate CP occurs in a topologically unstable gradient vector field on the verge of a structural change.
- Any type of gradient path always originates at a CP and terminates at another CP.

- The Poincaré–Hopf relationship, $n - b + r - c = 1$ enables a consistency check of a given pattern of CPs but cannot guarantee the completeness of this pattern.
- The quantity $\varepsilon = (\lambda_1/\lambda_2) - 1$ is called the ellipticity and measures the behaviour of the electron density in the plane tangential to the IAS at the BCP.

Further reading

Bader R. F. W., Anderson S. G. and Duke A. J. (1979). Quantum topology of molecular charge distributions. 1. *Journal of the American Chemical Society*, **101**, 1389–1395.

Collard K. and Hall G. G. (1977). Orthogonal trajectories of the electron density. *International Journal of Quantum Chemistry*, **12**, 623–637.

Chapter 6

Structural change

$$e^{+} + e^{-} \rightarrow 2\gamma$$
$$RCP + BCP \rightarrow *$$

6.1 Structure versus geometry

It is an infelicity that many textbooks do not distinguish between structure and geometry. They use these terms interchangeably and thereby confuse the fundamental insight which AIM offers by separating them and clearly defining what they mean. We will see that the distinction between structure and geometry is not a pedantic detail but rather a welcome product of a closer inspection of ρ's topology.

We use Figure 6.1 to clarify the difference between geometry and structure. This figure shows two different geometries of methanal that share the same structure. The *geometry* of a molecular system is defined as the list of nuclear coordinates, denoted collectively by **R**, determining the exact position in space of all nuclei. For our purposes it is not important to worry about the difference between 'absolute' geometry and 'internal' geometry.

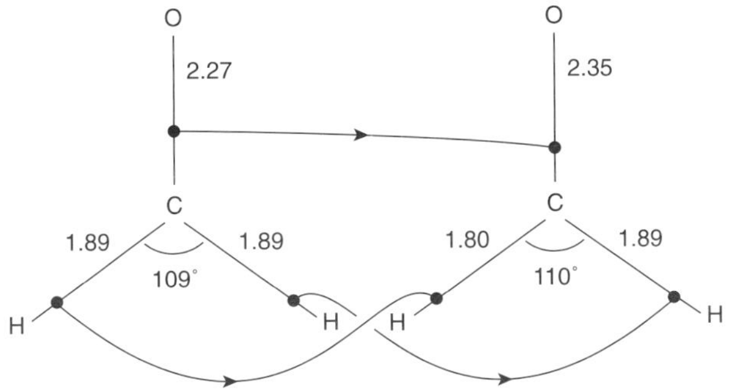

Figure 6.1 Schematic representation of two different geometries of methanal with the same MG. Although the actual position of the BCPs may differ, the BCPs link the same nuclei and therefore the topologically connectivity in the two molecules is the same. In view of this equivalence the two methanal molecules have an identical structure.

For example, if we discard rotation and translation of the molecule as a whole, the internal geometry of methanal is completely determined by the precise values of $3N - 6 = 6$ internal coordinates, where N is the number of nuclei. The two geometries of methanal shown in Figure 6.1 are obviously different because the bond lengths and angles differ slightly. It is important to realize that any *infinitesimal* change in coordinate values leads to a different geometry with coordinates $\mathbf{R}'$.

To understand the meaning of the concept of structure we look at the respective gradient vector field of both geometries of methanal, i.e. $\nabla\rho(\mathbf{r}; \mathbf{R})$ and $\nabla\rho(\mathbf{r}; \mathbf{R}')$ and ask ourselves whether they are topologically different. We observe from Figure 6.1 that the BCP between C and O of the left geometry can be mapped onto the corresponding BCP of the right geometry. The same observations apply to the other BCPs. In summary, the MGs of both geometries are equivalent in that their number and type of CPs is identical and in that the connectivity of the CPs is preserved. As a result, the gradient vector fields $\nabla\rho(\mathbf{r}; \mathbf{R})$ and $\nabla\rho(\mathbf{r}; \mathbf{R}')$ are *homeomorphic*, or topologically identical. A molecular *structure* is then defined as an *equivalence class* of MGs. The terms homeomorphic and equivalence class have a precise mathematical meaning which need not be given here to understand the main point made in this section. However, the interested reader can find this additional information in Box 6.1.

Box 6.1 A closer look

Equivalence and homeomorphism

A *relation* R in a set A is actually a subset of the product space $A \times A$ (all possible couples (a_i, a_j)). For example, A is a set of n geometries $\{G\}$ that a given molecule can adopt and R is the prescription 'has the same MG as'. In this example R is simply the set of n^2 couples (G_i, G_j), meaning that G_i has the same MG as G_j. Now, R is called an *equivalence relation* if and only if R satisfies the following three conditions:

1. R is *reflexive*, i.e. for every $a \in A$, $(a, a) \in R$.
2. R is *symmetric*, i.e. if $(a, b) \in R$ then $(b, a) \in R$.
3. R is *transitive*, i.e. if $(a, b) \in R$ and $(b, c) \in R$, then $(a, c) \in R$.

It is trivial to check that the relation 'has the same MG as' is an equivalence relation. A most useful mathematical concept with direct chemical meaning is the *equivalence class*. If R is an equivalence relation in A, then the equivalence class of any element $a \in A$, denoted by $[a]$ is the set of elements to which a is related or $[a] = \{x : (a, x) \in R\}$. In our example, this means that if we pick a geometry G_i then its equivalence class $[G_i]$ is the set of all geometries that have the same MG, which is nothing else than the structure of G_i.

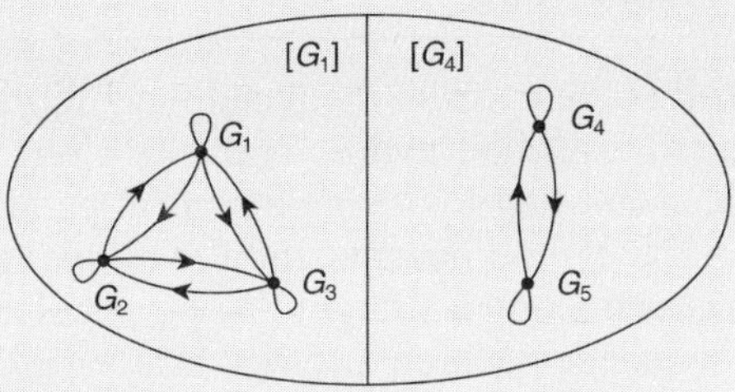

Figure 1 A schematic representation of a set of five geometries partitioned into two equivalence classes.

Omitting some further mathematical details, one can rigorously prove that the set of equivalence classes $\{[a]\}$ is a *partition* of A. A partition is a set of (1) non-empty, (2) pairwise disjoint sets such that (3) each $a \in A$ belongs to some set. This idea is illustrated in Figure 1, where A is a set of five geometries $\{G_1, G_2, G_3, G_4, G_5\}$ of a single molecule, partitioned into two non-overlapping equivalence classes. Each class is labelled by a representative of the class, for example $[G_1] = [G_2] = [G_3]$. Translated into chemical terms this example tells us that we have discovered two structures of the molecule, one represented by three possible geometries and the other by two geometries.

The level of this book prevents an elaborate mathematical explanation of a *homeomorphism* but one should realize the following: two topological spaces X and Y are called *homeomorphic* or *topologically equivalent* if there exists a bijective (i.e. one-to-one and onto) mapping $f : X \rightarrow Y$ such that f and its inverse f^{-1} are continuous. The function f itself is then called a *homeomorphism*. In more flowery terms this means that if two objects, such as MGs can be 'plastically deformed' in a *reversible* way then they are topologically equivalent. It is often mentioned in topology textbooks that all deformations are allowed except tearing, shearing, cutting or gluing. Clearly, two MGs belonging to different structures cannot be continuously deformed into each other without creating or destroying singularities such as CPs.

Homeomorphisms do not necessarily occur between gradient vector fields of the same function (such as ρ) but can also exist between vector fields of a different function. For example, it has been observed that $\nabla\rho$ is homeomorphic to ∇V where V is the classical electrostatic potential of a molecule.

Although moderate vibrations in a molecule change the configuration of the nuclei, and therefore the geometry, they typically leave the equilibrium structure intact. However, floppy molecules or more rigid molecules in highly excited vibrational states may fluctuate between several structures all of which can be decisively distinguished using AIM. A structure is in fact an open neighbourhood of nuclear configuration space and changes in structure turn out to be abrupt and discontinuous processes as shown in the next section.

6.2 Bifurcation mechanism

What would happen if we start from methanal's equilibrium geometry and pull the C=O unit away from the hydrogen nuclei along the C_2 axis? This change can easily be quantified by a single control parameter u defined as the distance between the carbon nucleus and the midpoint between the two hydrogen nuclei measured along a straight line perpendicular to the C_2 axis. Figure 6.2 shows a sequence of three different structures obtained by fixing u at 1.10, 2.46 and 2.48 au for (a), (b) and (c) respectively. The first geometry generates an MG which belongs to the equilibrium structure (a) and which is stable for quite large distortions of the nuclear skeleton. A drastic change in geometry, however, produces a ring structure (c), which is again stable under moderate excursions of the nuclei. On the path from (a) to (c) we encounter the topologically unstable structure (b) which is energetically *stable*, i.e. it is *not* an (energy) transition state. Figure 6.2 clearly illustrates that the geometries in (a) and (c) have different structures because their gradient vector fields $\nabla\rho(\mathbf{r};\, \mathbf{R}_a)$ and $\nabla\rho(\mathbf{r};\, \mathbf{R}_c)$ cannot be mapped onto each another in a one-to-one way. Indeed, although the two C–H BCPs can still be matched, the H–H BCP does not occur in structure (a), nor does the RCP. The unstable structure (b) is also different from (a) and (c) because it uniquely contains a degenerate CP of rank 2 between the two hydrogens, denoted by (2, 0). It is at this point that a new CP *abruptly* appears between the hydrogen nuclei thereby marking the formation of the H–H bond. The reaction path followed expresses in (b) that methanal is being dissociated into molecular hydrogen and carbon monoxide.

Figure 6.3 zooms in on another aspect of the sequence shown in Figure 6.2: the electron density profile along the C_2 axis for structures (a), (b) and (c). It is easier to discuss the profiles from (c) to (a). In (c) the RCP and BCP, both lying on the C_2 axis, show up as extrema in the ρ profile because the definition of a CP, $\nabla\rho = \mathbf{0}$, ensures that the derivative viewed in any direction vanishes. The BCP lies on the AIL between the two hydrogen nuclei and is the centre of the H|H IAS which contains one gradient path originating at the RCP. Following the convention that $\lambda_1 \le \lambda_2 \le \lambda_3$, it can be proven that for both RCP and BCP the eigenvector $\mathbf{u}_2$ lies along the C_2 axis. Both the negative curvature at the BCP and the positive curvature at the RCP correspond to the respective λ_2 eigenvalues at each CP. The pair of eigenvalues (λ_1, λ_3) has the sign pattern (–, +) at the BCP as well as at the RCP and can be ignored in the present discussion. As the RCP and the BCP come nearer, their respective curvatures in the $\mathbf{u}_2$ direction become flatter. The only way the BCP's negative curvature ($\lambda_{2,\mathrm{BCP}} < 0$) and the RCP's positive curvature ($\lambda_{2,\mathrm{RCP}} > 0$) can become identical upon critical point coalescence is by attaining the value zero. This is exactly what happens at (b), because here the RCP and the BCP have merged into a (2, 0) degenerate CP which manifestly has one zero curvature ($\lambda_{2,\,(2,0)} = 0$). At this CP we find a singularity, called a *point of inflexion*, where the curve is locally flat since

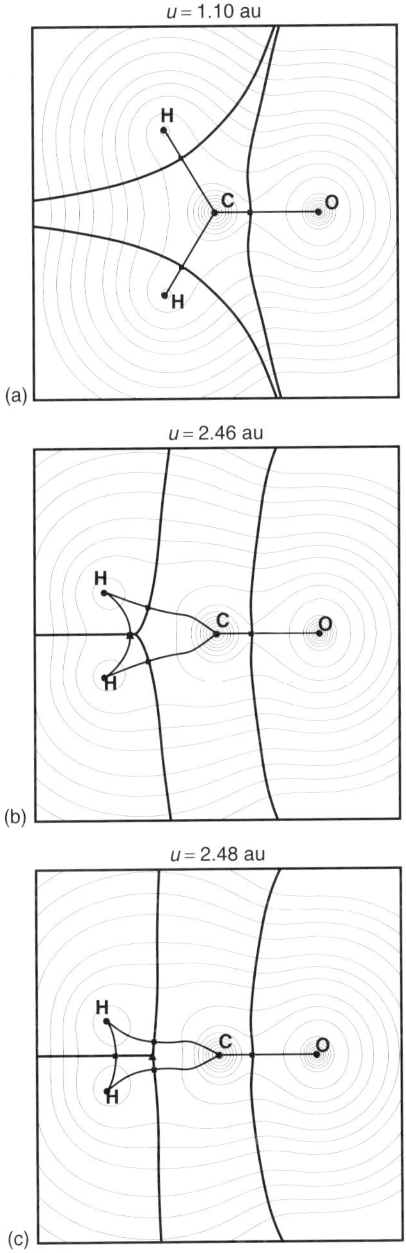

Figure 6.2 Superposition of the electron density contour map, the MGs and the IASs of three different structures of methanal. Structure (a) is the equilibrium structure, (b) is a *topologically* (not energetically) unstable intermediate and (c) corresponds to a stable ring intermediate. The sequence of structures is obtained by systematically varying the control parameter u from 1.10 au via 2.46 au to 2.48 au, in (a), (b) and (c) respectively.

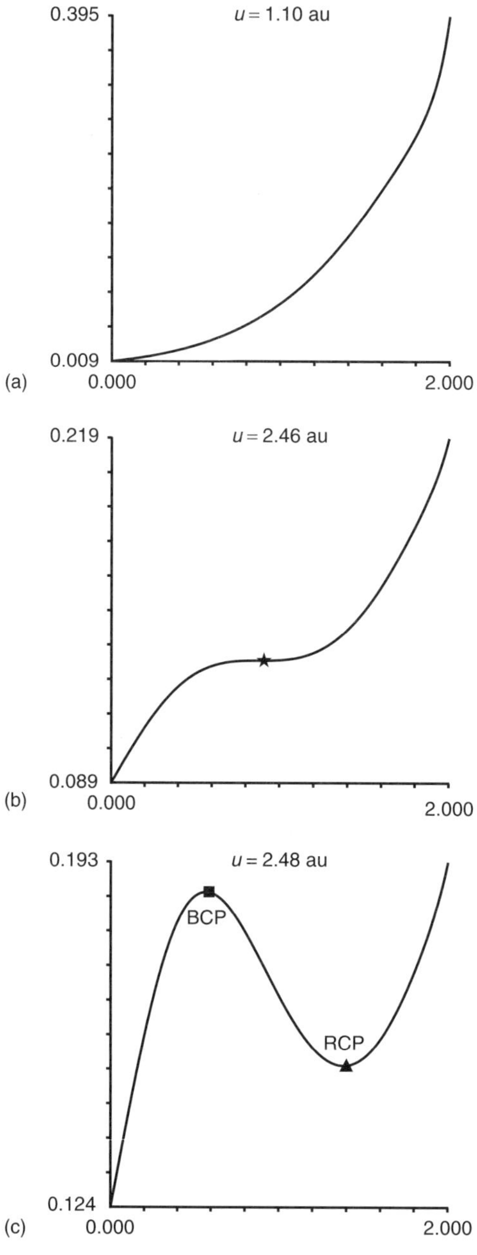

Figure 6.3 One-dimensional curves representing the electron density (in atomic units, au) versus a displacement along the C_2 symmetry axis (au) in the dissociation process of methanal. The different structures are obtained by parametrically varying the control parameter u. The present sequence, marked by (a), (b) and (c), corresponds exactly to the one given in Figure 6.2. The star in structure (b) marks the unstable CP, which is a point of inflexion.

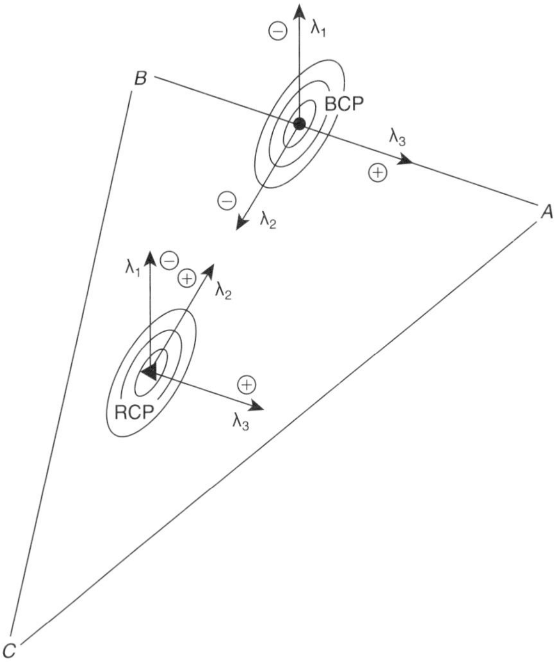

Figure 6.4 A schematic representation of the local curvatures (sign marked by ⊕ and ⊖ in the merging process of a BCP (black dot) and an RCP (solid triangle) in a general ring system constituting three arbitrary nuclei *A*, *B* and *C*. Ellipsoid contour lines represent the behaviour of the density near the CPs. As the two CPs approach, their respective λ_2 eigenvalues get nearer to zero until they perfectly vanish when the RCP and BCP coalesce. Note that λ_1 of the RCP is the only negative eigenvalue and describes the ring line, while the only positive curvature at the BCP describes the AIL linking *A* and *B*.

the second derivative vanishes. The structure appearing in (b) is termed a *bifurcation structure* and the point in nuclear configuration space is referred to as the *bifurcation point*. The name 'bifurcation' is obviously inspired by the phenomenon that a degenerate CP 'splits up' in two complementary objects, the RCP and the BCP. Finally, when the C=O entity is pushed further into the H_2 unit, all CPs along the C_2 symmetry axis disappear and a profile such as that in (a) ensues.

The CP merging process is illustrated in Figure 6.4 in three dimensions for a general three-membered ring system. Each CP is endowed with a local (eigenvector) axis system with associated local curvatures expressed by the eigenvalues of the Hessian. As mentioned before, the $\mathbf{u}_2$ eigenvectors of the two CPs are aligned and mark the direction of the soft curvature exhibited by the ellipsoid ρ contour diagrams at each CP. As the CPs approach, their curvatures become so soft that they disappear altogether when they merge into a (2, 0) CP. The ellipticity ε of the BCP, which we defined before as

$\lambda_1/\lambda_2 - 1$, becomes infinite as the BCP merges with the RCP. A high ellipticity (for example, more than 5) indicates that a bond is structurally unstable in the sense that the concomitant BCP may annihilate upon uniting with a nearby RCP.

We have discussed essential features of the so-called *bifurcation mechanism*, one possible way to induce a structural change. It should be mentioned again that a structural change is an abrupt and discontinuous process, a direct consequence of the unambiguous formulation of molecular graphs. In other words, an MG is a precise scheme of connectivity obeying binary logic, i.e. two nuclei are linked by an AIL or they are not, there are no 'in-betweens' or gradations. As a result, if a reaction transforms molecule *A* into molecule *B* which possesses a different structure, then the transition must have involved at least one *discontinuous* structural change.

It will become clear in sections 6.4 and 6.5 how the bifurcation mechanism fits into the broader picture of structural change. At this stage we just point out that it occurs in more instances than the reader may expect. For example, an amino acid in a peptide environment may occur in the so-called C_5 conformation (see Figure 10.7a). The (ψ, ϕ) torsion angles of this conformation enable the formation of an intramolecular hydrogen bond including five atoms. However, this is a topologically unstable conformation, as gathered from the small distance between the BCP and RCP and from the high ellipticity of the hydrogen bond BCP. It is one of the useful applications of AIM that one can assess the stability of a molecular conformation based on a *single* electron density. For example, when confronted with a multiple hydrogen bond scheme one can predict via a single snapshot of a dynamic sequence of conformations which bond(s) will be ruptured first.

6.3 Conflict mechanism

Let us start from the methanal structure shown in Figure 6.2(c), where the control parameter u was set to 2.48 au. Pulling the C=O entity further to the right makes the C–H AILs come closer together until the two C–H BCPs and the RCP all merge and leave one BCP behind exactly on the C_2 axis. This new structure is illustrated in Figure 6.5(b) where $u = 2.50$ au. Although this process appears to be a fundamentally new mechanism of structural change, it is not. Instead, it constitutes the simultaneous occurrence of two bifurcation mechanisms, i.e. one where the upper C–H BCP coalesces with the RCP and one where the lower C–H BCP coalesces with the same RCP. It will become clear in section 6.4 (Figure 6.7) how this structural change relates to the other type.

In any event, Figure 6.5(b) is our starting point to understand a new type of structural change called the *conflict mechanism*. In this figure there is a BCP lying on the C_2 axis, which links the carbon nucleus and the BCP

Plate section

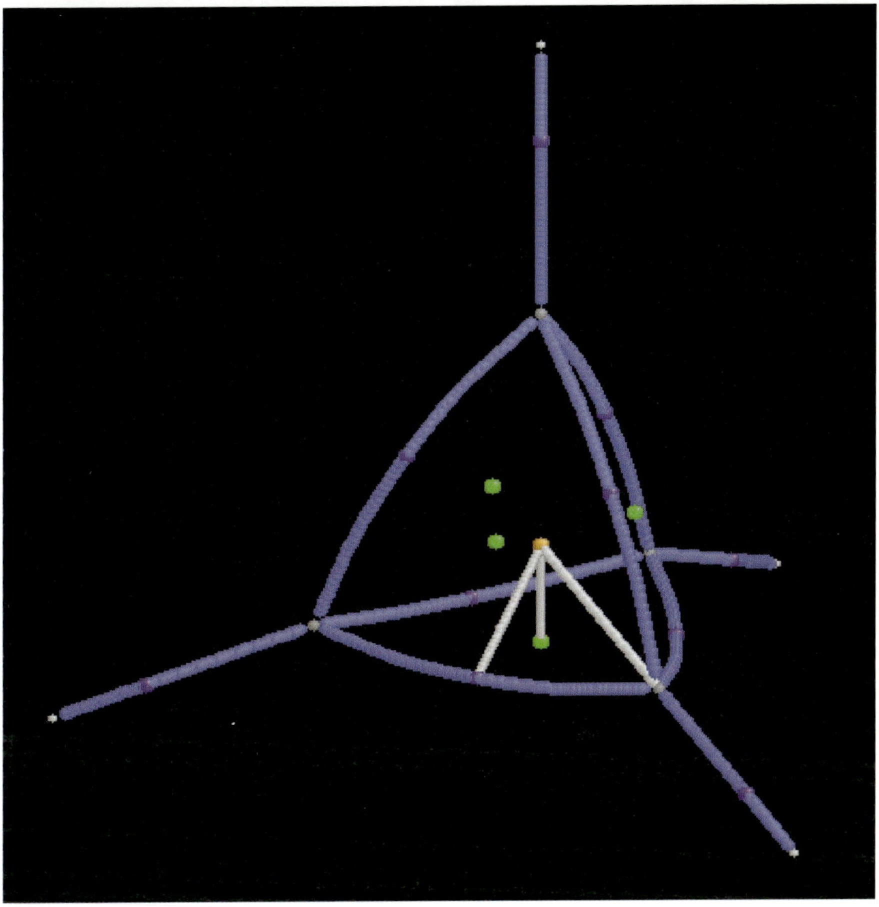

Figure 5.2 A 3D view of tetrahedrane, C_4H_4, which consists of a tetrahedron of carbons (grey), each bonded to a hydrogen (white). The CCP (yellow) appears at the centre of the tetrahedron. The ten BCPs are marked in purple and the BPs are marked in blue. The four RCPs are shown in green, one for each tetrahedron face. Three gradient paths are shown in white, each originating at the CCP and terminating at a nucleus, a BCP and a RCP respectively.

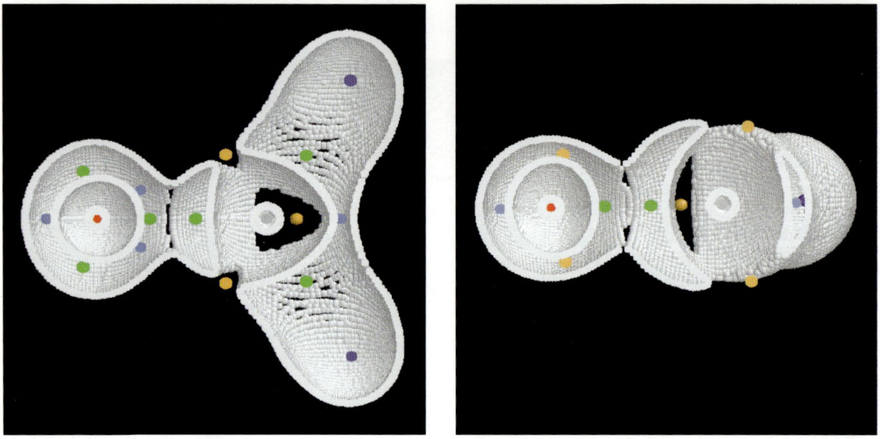

(a) (b)

Figure 8.4 (a) Representation of the zero envelope of $\nabla^2\rho$ of methanal in the presence of the atomic graphs of oxygen and carbon. The two green dots on the left are the non-bonding (3, –3) charge concentrations. The next green dot to the right is the bonding (3, –3) of oxygen and the next one is the bonding (3, –3) of carbon. The pair of green dots on the utmost right are the two bonding (3, –3) maxima of carbon's atomic graph involved in the bonds with hydrogen (in purple). The yellow dots are three of the four (3, +1) minima of carbon's atomic graph. The fourth (3, +1) can be obtained by mirroring the right yellow dot in the molecular symmetry plane. The three blue dots on the left constitute the full set of (3, –1) saddle points of oxygen's atomic graph. The blue dot on the right is one of the five (3, –1) saddle of the atomic graph of carbon. (b) Another slice through the topological object represented in part (a), made by the plane through C and O bisecting the two hydrogens. The same colour code as in (a) is used.

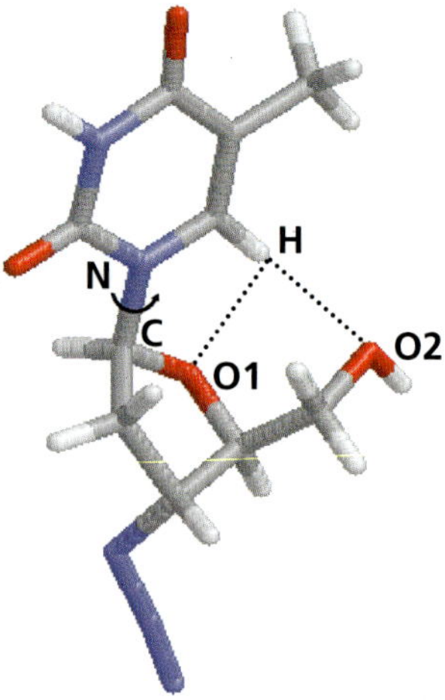

Figure 10.5 A stick representation of AZT. The bifurcated hydrogen bond is marked by dotted lines. The C–H · · · O1 bond is weaker than the C–H · · · O2 bond. The glycosyl torsion angle is centred on the C–N, and marked with an arrow. In this unusual conformation of AZT, which has been detected crystallographically, this torsion angle is near zero. (From Koch and Popelier, 1995. Copyright © American Chemical Society; reprinted by permission.)

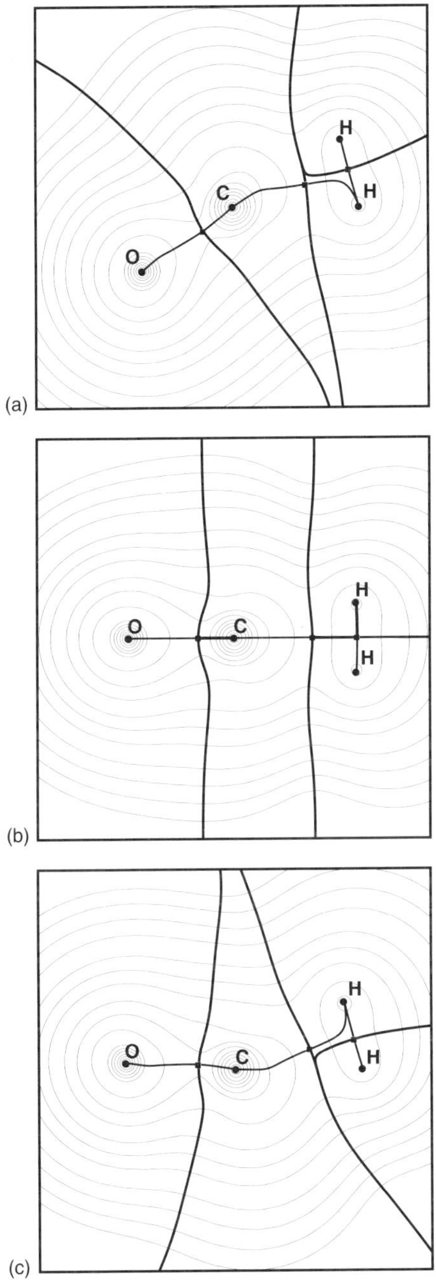

Figure 6.5 The conflict mechanism illustrated for three different structures of methanal. The unstable structure of (b) can develop either way, i.e. towards the stable structure (a) or (c). A tiny displacement of one of the nuclei is sufficient to disrupt the unstable conflict structure (b) but here we show a fairly major geometrical change.

between the hydrogens. This situation is remarkable because normally a BCP links two nuclei rather than a nucleus and another BCP. It is as if the BCP linked to the carbon nucleus 'is undecided' about which of the two hydrogen nuclei it should link. Strictly speaking, the carbon is bonded to the two hydrogens at the same time. Another view is that the two hydrogen nuclei are in competition to be linked to this BCP, or in other words they are in *conflict*. Figure 6.5(a) shows a stable structure in which the conflict has been resolved in favour of one hydrogen nucleus, and Figure 6.5(c) shows resolution in favour of the other one. Note that in these figures the geometry changes that induced the conflict's resolution are larger than required. In fact, an infinitesimal geometrical change is sufficient to make the *conflict structure* flip to one or the other structure. It is characteristic of a conflict structure that all CPs in its MG are of rank 3 and therefore stable while the graph itself is unstable.

In summary, this is a fundamentally different kind of topological instability from the one encountered in bifurcation structures. Normally an IAS contains one central BCP at which all the IAS gradient paths terminate. Here, the IAS between the hydrogens includes a gradient path *originating* at a BCP, other than the usual central BCP. Any motion of the nuclei which causes this gradient path to leave the IAS will make the gradient path abruptly switch to one or the other attracting hydrogen nuclei. Thus the conflict structure is unstable with respect to *some* nuclear motion but not all. Indeed, a further increase of the control parameter u, thereby preserving perfect C_2 symmetry, will sustain the conflict structure.

The thermal isomerization of HCN to CNH (Bader, 1990, Figure 3.6) is a simple example of a reaction encompassing one conflict structure. The hydrogen nucleus migrates from one position, being bonded to the carbon nucleus, to another position, being bonded to the nitrogen nucleus. This reaction can be followed in both ρ and the energy by computing a succession of wave functions, one for each nuclear configuration. As the hydrogen nucleus migrates over the carbon nucleus, it suddenly becomes connected via a bond path to the nitrogen nucleus. This unstable conflict structure can be computed to great precision and happens to correspond to the reaction transition state. This perception of energy and electron density in transition states going hand in hand is, unfortunately, not generally observed. Such remarkable correspondence warrants further research.

The two mechanisms discussed here are the only ones that can describe a change in the topological structure of ρ. For sake of illustration we now step back for a moment and concentrate on the overall picture the conflict and bifurcation mechanisms offer in the migration of the CH_3^+ group over the ethene moiety (C_2H_4) in the carbonium ion $C_3H_7^+$. Figure 6.6(a) shows the conflict type in which the two ethene carbons are competing for the methyl carbon. Figure 6.6(b) involves two stable mirroring structures each connected to a central ring intermediate via two unstable mirroring structures each having a degenerate (2, 0) CP.

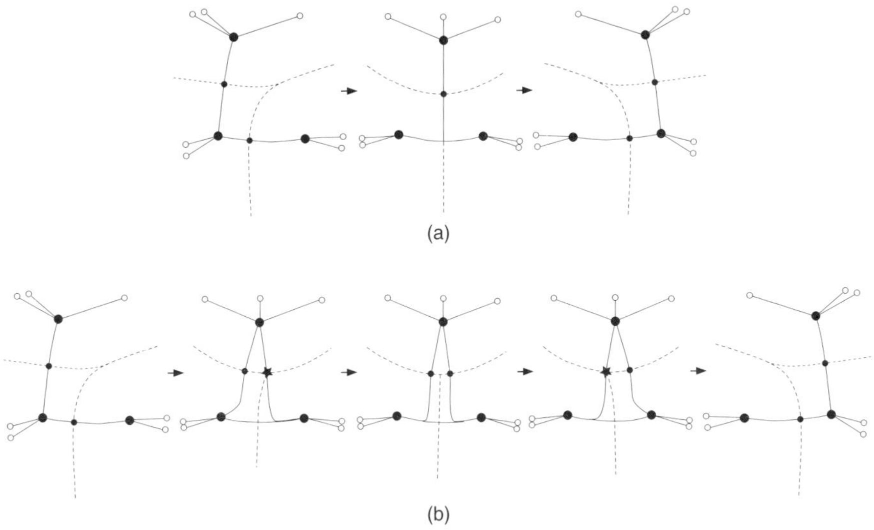

Figure 6.6 Molecular graphs (determined by calculations not including electron correlation) illustrating structural changes in the molecule $C_3H_7^+$. The sequence shown in (a) shows the conflict mechanism and the one in (b) the bifurcation mechanism. In both mechanisms a CH_3^+ group is migrating over the ethene moiety (C_2H_4). The hydrogen nuclei are denoted by open circles, the carbon nuclei by full circles and the C|C IASs by dashed curves. The degenerate CP is marked by a star. (From Bader *et al.*, 1981. Reproduced by permission.)

6.4 Structure diagram

A quantitative, detailed description of changes in the MG is not very useful. Instead, we wish to identify in these changes the notion of structure, an important invariant property which can be described precisely in the language of mathematics. In this and the next section we increase the level of abstraction in order to put the only two possible mechanisms of structural change on a common footing.

In the example of methanal, the changes in the MGs were generated predominately by the control parameter u. However, this parameter alone was not sufficient to distort the conflict structure into a stable structure in which a BCP connected the carbon nucleus and *either* hydrogen nucleus. For that purpose we push the oxygen nucleus (slightly) off the symmetry axis, a change which can be described by a second independent control parameter v. These two control parameters form a 2D control space. In general, the M internal coordinates governing the positions of the nuclei make up an M-dimensional *control space*. However, a control parameter can be associated with combined motion of nuclei. The discussion of structure requires another space however, called the *behaviour space*. This is the space wherein the behaviour of the system is determined and observed through the properties

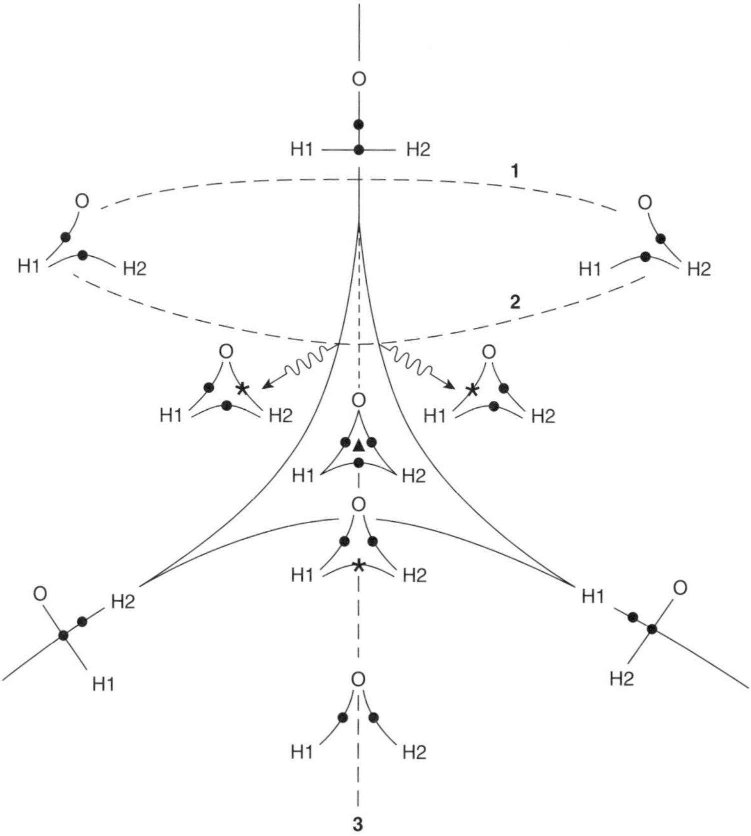

Figure 6.7 The structure diagram of water. The full lines, marking the set of catastrophe points, consist of an inner hypocycloid (curved triangle) and attached three semi-axes. The former is the locus of the bifurcation structures and the latter of the conflict structures. All other points in this diagram are regular points. Three different paths (1, 2, 3) representing a sequence of structural changes, are marked by dashed lines. Path 1 links two structures of water via the conflict structure associated with the vertical semi-axis. Path 2 links the same structures via two bifurcation structures on opposite branches of the hypocycloid (* marks the (2, 0) CP) and one stable intermediate ring structure inside the hypocycloid. Path 3 links an open water structure (bottom) via the third possible bifurcation structure, [O–(H1–*–H2)], and the ring intermediate to a conflict structure (top). (From Bader *et al.*, 1981. Reproduced by permission.)

of the electron density. Since the electron density exists in our familiar and real 3D space, this space is the behaviour space for ρ's topology.

Figure 6.7 shows the *structure diagram* of the three-nucleus system water. A structure diagram is a picture which summarizes all possible structures and all mechanisms of structural change for a given chemical system. This diagram can be used for a better understanding of the previous discussion of the structural changes observed in our methanal example. A structure diagram contains two types of point: *regular points* and *catastrophe points*. The former lie in a structurally stable region; the latter lie on the structurally unstable boundaries of these stable regions. The structure diagram shown in Figure 6.7 has four structurally stable regions, symbolically labelled by [H1–O–H2], [O–H1–H2], [O–H2–H1] and [H1–O–H2]. The first structure is the ring structure inside the (central) hypocycloid (or curved triangle), whereas the last one is the well known structure of water. Each regular point corresponds to a particular geometry controlled by two parameters u and v. The set of catastrophe points falls apart in two different subsets: the conflict structures and the bifurcation structures. The conflict structures on the three semi-axes touching the hypocycloid and can be conveniently labelled as [O–(H1–H2)], [H2–(O–H1)] and [H1–(O–H2)]. The bifurcation structures lie on the boundaries of the hypocycloid and are tagged [O–(H1–∗–H2)], [H2–(O–∗–H1)] and [H1–(O–∗–H2)].

In Figure 6.7 we marked three journeys in the structure diagram, each represented by a dashed line. The first path is the simplest and links two stable structures (regular points) via an obvious conflict structure. This path also describes the reaction path of one type of CH_3^+ migration in $C_3H_7^+$ (Figure 6.6a) as well as the conflict situation encountered in methanal's dissociation. The second path meets two bifurcation structures flanking a stable ring intermediate. We encountered this path in Figure 6.6(b) where the alternative, more involved mechanism for CH_3^+ migration was illustrated. The third and final path consorts with the very first example of structural change we discussed in this chapter, the bifurcation mechanism of methanal of Figure 6.2, provided the O is replaced by C=O. The bottom structure then corresponds to the equilibrium structure of methanal. As we move along this path from the bottom upwards, we first encounter the expected and extensively described structure containing the (2, 0) CP (Figure 6.2b). Consequently we travel inside the stable region of the ring intermediate (Figure 6.2c). As we move up further into the upper corner of the hypocycloid the O–H1 and O–H2 AILs merge as the structure turns into the upper conflict structure. It should be emphasized again that this conflict structure was obtained via the third path, which involves only bifurcation mechanisms, rather than a conflict mechanism.

A further abstraction of the present considerations would invoke a mathematical branch, very popular in the 1970s and pioneered by the French mathematician Thom, called *catastrophe theory*. It was realized soon after the publication of his seminal book that catastrophe theory could be used to describe changes in molecular structure. The theory has a much wider range of application because its purpose is to explain sudden, discontinuous changes in the behaviour of a system which are ubiquitous throughout Nature. These changes are smooth and continuous and usually surface through an alteration in form. In biological systems this is known as morphogenesis for example. Structural changes in molecules such as H_3^+ and [1.1.1]propellane (which involves the formation of a cage structure) can all be put on a common footing using catastrophe theory introducing the unfolding of the so-called elliptic umbilic.

Summary

- Geometry and structure are two entirely different things. The former is determined by the exact nuclear positions, the latter is a connectivity scheme governed by the topology of ρ.
- If two MGs can be continuously mapped into each other (one-by-one) they are homeomorphic or topologically equivalent.
- Molecular structure is an equivalence class of the relation 'has the same MG' operating on the set of all molecular geometries.
- There are only two types of structural change possible: the bifurcation mechanism and the conflict mechanism, each possessing a corresponding unstable structure.
- In the bifurcation mechanism a BCP and a RCP annihilate and leave a degenerate (2, 0) CP behind. The curvature associated with the second eigenvector of the Hessian vanishes and the ellipticity of the BCP becomes infinite.
- In the conflict mechanism a conflict structure occurs in which all CPs have rank 3 but the MG is unstable. The conflict consists of two nuclei competing for a single BCP and can be resolved by an infinitesimal distortion of the conflict geometry.
- Parameters controlling geometry changes constitute the control space affecting the topology of ρ which exists in a 3D behaviour space.
- A structure diagram puts all possible structures and structural changes on a common denominator. It contains structurally stable (regular points) and unstable regions (catastrophe points).
- Catastrophe theory is a branch of mathematics which classifies and describes abrupt changes. It can be used to obtain a full and precise understanding of changes in chemical structure via an AIM topological analysis of ρ.
- Remarkable examples are known of structural changes where the energy transition state coincides with the unstable structure in the electron density. Unfortunately this link does not hold in general.

References

Bader R. F. W. (1990). *Atoms in Molecules: A Quantum Theory*. Oxford: Clarendon.

Further reading

Bader R. F. W. (1985). Atoms in molecules. *Accounts of Chemical Research*, **18**, 9–15.

Bader R. F. W., Nguen-Dang T. T. and Tal Y. (1981). A topological theory of molecular structure. *Reports on Progress in Physics*, **44**, 894–948.

Thom R. (1975). *Structural Stability and Morphogenesis*. New York: W. A. Benjamin.

Chapter 7

The quantum atom

If an atom is a state
Then this is its federal law

7.1 The rules of the game

The reader should be reassured that the title of this chapter does *not* mean that a new type of atom will be introduced. Instead, we will learn here that the atom that we defined previously as a purely topological object is also firmly rooted in quantum mechanics (Bader and Nguyen-Dang, 1981). This is exactly what one would expect from a theory that acts as a bridge between quantum mechanics and chemistry. A recurring theme in the formulation of atomic subspace quantum mechanics is that finite boundaries create 'surface terms' which do not necessarily vanish. Therefore the *subspace* quantum mechanical expressions corresponding to those for the total (molecular) system are more general and hence more complicated. However, if the subspace is properly chosen, certain terms vanish. Such a proper choice is exactly the topological atom that we introduced in Chapter 3, thereby making it even more desirable than purely on the grounds of topological insight. In summary, our atoms are special objects in both a topological and a quantum mechanical sense.

As stated in Chapter 1, the founders of quantum mechanics have left us with a consistent and powerful formalism, i.e. a set of rules that we have to apply systematically to predict the physics of a quantum system. The benefit of these rules is that one can 'switch off' and perform the calculations almost mechanically. Applying these rules may look difficult but will be rewarded by the beauty of the end result. Note that we use atomic units (au) throughout this chapter, i.e. fundamental constants such as m_e and $\hbar$ are set to unity.

We invite the reader to digest the formulae in the text, to contemplate their meaning and to follow the logic of the derivations even if the workings are not given in detail. This chapter is intended to be self-contained given a basic working knowledge of quantum mechanics and familiarity with typical equation manipulations. To see the wood for the trees we have inserted many boxes with brief reviews and details of key concepts necessary to follow the main thread. An important piece of mathematics called vector calculus, is given in Box 7.1. It provides us with Gauss' divergence theorem, which is ubiquitously applied throughout derivations of quantum mechanical equations for atoms. Basically it equates a volume integral to a surface integral. This means that atomic properties, which are defined as volume integrals over the atomic basin, can be linked to a different physical property over its surface, as we shall see below.

Box 7.1 A closer look

Some vector calculus

Vector calculus is a branch of mathematics inspired by fluid dynamics and electrostatics. We need only a few key elements to proceed in our quantum mechanical derivations. The first important element is *Gauss' divergence theorem*, which establishes a link between a volume integral and a surface integral. Let Ω be a region in space bounded by a surface S, $\mathbf{V}$ be a vector field defined in Ω and $\mathbf{n}$ be the normal to the surface S, then:

$$\iiint_{\Omega} dV \nabla \cdot \mathbf{V} = \iint_{S} dS\ \mathbf{V} \cdot \mathbf{n}$$

The proof essentially uses the fact that integration and differentiation are inverse operators. Indeed, one of the integrations in the triple (volume) integral annihilates the differentiation of the ∇ operator, thereby leaving a double (surface) integral.

The surface integral is in fact the *flux* of the vector field $\mathbf{V}$. Note that the concept of flux contains two 'components': an orientation, taken into account by the dot product, and a summation of infinitesimal contributions, expressed by the surface integral. In that sense it is useful to think of flux as 'how much wind a sail catches' depending on its direction in the wind and its size. However, we do not need a stream of moving particles to define and use the concept of a flux. The vector field in question may well be an electrostatic field or a gravitational force field. In fact, when applied to those two fields, Gauss' divergence theorem turns out to be very useful. For example, it follows from this theorem that the force produced by a solid sphere of matter is the same at the sphere's surface as it would be if all the matter were concentrated at the centre. This fact, which follows from our theorem, used to worry Newton so much he did not make public his theory of gravitation for many years.

The integrand of the volume integral is in fact the *divergence* of the vector field $\mathbf{V}$ or:

$$\text{div}\ \mathbf{V} = \frac{\partial V_x}{\partial x} + \frac{\partial V_y}{\partial y} + \frac{\partial V_z}{\partial z}$$

To understand the geometric meaning of divergence (Marsden and Tromba, 1983, p.183ff) we imagine an infinitesimal volume element somewhere on a gradient path (also known as flow line or streamline) of the vector field **V**. Now move this little element along the gradient path and measure how its volume changes with respect to time. The more dramatic this change, the larger the divergence at a given point, and if it is negative then this volume shrinks. This interpretation (which can be expressed as a non-trivial theorem) points towards the divergence actually describing compressibility and 'density' changes. For example, if the divergence of a fluid is everywhere zero we say it is incompressible. It is instructive to define the *Laplacian* (denoted ∇^2) in this context provided $\mathbf{V} = \nabla f$ where f is some scalar 3D function such as the electron density. The Laplacian of f, $\nabla^2 f$, is then defined as the divergence of ∇f or:

$$\nabla^2 f = \nabla \cdot \nabla f$$
$$= \frac{\partial^2 f_x}{\partial x^2} + \frac{\partial^2 f_y}{\partial y^2} + \frac{\partial^2 f_z}{\partial z^2}$$

As explained in Chapter 8, the Laplacian is a measure of local depletion or concentration of f.

A useful corollary of the divergence theorem is one of *Green's identities*. The proof involves a clever combination of a divergence product rule, i.e. $\nabla \cdot (f\nabla g) = \nabla f \cdot \nabla g + f\nabla \cdot \nabla g$ followed by the equation of a volume integral to a surface integral resulting in:

$$\iiint_\Omega \mathrm{d}V(f\nabla^2 g - g\nabla^2 f) = \iint_S \mathrm{d}S(f\nabla g - g\nabla f)\cdot \mathbf{n}$$

A simple but very important application of this equation is that $\hat{H}$ is not Hermitian for a subspace. Indeed,

$$\int_\Omega \mathrm{d}\tau\psi^*(\hat{H}\psi) \neq \int_\Omega \mathrm{d}\tau\psi(\hat{H}\psi)^*$$

which can be proven by computing the difference between the right- and left-hand sides of this equation. Using the fact that $\hat{H} = -\frac{1}{2}\nabla^2 + \hat{V}$ we obtain that

$$\iiint_\Omega \mathrm{d}V[\psi^*(\nabla^2\psi) - \psi(\nabla^2\psi)^*] = \iint_S \mathrm{d}S(\psi^*\nabla\psi - \psi\nabla\psi^*)\cdot \mathbf{n}$$

which is different from zero except for full space, because for a 'bounding' surface at infinity, ψ and its derivatives vanish.

A direct and important application of Gauss' divergence theorem is that the volume integral over the atomic basin Ω of the Laplacian of ρ, $\nabla^2\rho$, vanishes or

$$\iiint_\Omega \mathrm{d}V\nabla^2\rho = 0 \tag{7.1}$$

Indeed, if we take the vector field **V** to be $\nabla\rho$ and apply the divergence theorem then:

$$\iiint_{\Omega} dV \nabla^2\rho = \iiint_{\Omega} dV \nabla \cdot \nabla\rho = \iint_{S} dS \nabla\rho \cdot \mathbf{n}$$

The last equality above results from applying Gauss' theorem. The proof of equation (7.1) is completed by noting that $\iint_{S} dS \nabla\rho \cdot \mathbf{n} = 0$ because S is an IAS for which $\nabla\rho \cdot \mathbf{n} = 0$. In fact atoms extending to infinity are bounded at infinity by a surface for which $\nabla\rho \cdot \mathbf{n}$ also equals zero since $\nabla\rho$ vanishes there. Note that the extent to which the equation $\iiint_{\Omega} dV \nabla^2\rho = 0$ holds is a measure for the quality of an atomic integration (see section 3.4). In other words, any deviations from zero by this integral can be considered as an integration error.

A second important application of Gauss' divergence theorem, via a corollary known as one of Green's identities (Box 7.1), is that an operator which is Hermitian (Box 7.3) for total space is *not* Hermitian for a subspace. This has serious consequences for the formulation of subspace quantum mechanical equations as we will see below.

7.2 The kinetic energy of an atom

Would it not be remarkable if one could rigorously define the energy of an atom *inside a molecule*? AIM makes this possible. However, in this section we discuss just the necessary prerequisites to reach this goal. Here we demonstrate how we can define the *kinetic* rather than the total energy of a single atom inside a molecule. It turns out that there is more than one way to define the kinetic energy *density* at a point. The word 'density' implies that the kinetic energy is evaluated *locally* at a point just like the electron density. Integration of a property density yields the (integrated) property over a given volume, which is the kinetic energy in the present case. One type of kinetic energy density is denoted by $K(\mathbf{r})$, and another by $G(\mathbf{r})$ where the **r** are the coordinates of a position vector. These two types of kinetic energy density are defined as follows:

$$K(\mathbf{r}) = -\frac{1}{4} N \int d\tau' [\psi^* \nabla^2 \psi + \psi \nabla^2 \psi^*] \tag{7.2}$$

$$G(\mathbf{r}) = \frac{1}{2} N \int d\tau' \nabla\psi^* \cdot \nabla\psi \tag{7.3}$$

where $N\int d\tau'$ summarizes the 'one electron integration mode' introduced in Box 1.1. The reader is reminded that this integration prescription extracts a *one-electron property* from the full electronic wave function ψ because we integrate over all coordinates of the N electrons except the spatial coordinates of one electron. Applying the differentiation product rule of ∇^2 operating on $\psi^*\psi$ yields $\psi^*\nabla^2\psi + \psi\nabla^2\psi^* + 2\nabla\psi \cdot \nabla\psi^*$. After $N\int d\tau'$ integration, this expression leads to the following relation between $K(\mathbf{r})$ and $G(\mathbf{r})$:

$$K(\mathbf{r}) = G(\mathbf{r}) - \frac{1}{4}\nabla^2\rho(\mathbf{r}) \qquad (7.4)$$

where we have used that $\rho(\mathbf{r}) = N\int d\tau'\psi^*\psi$.

If we integrate equation (7.4) over full space then we obtain that K(full space) = G(full space) since the last term vanishes using equation (7.1). Thus the kinetic energy of a molecule (i.e. the system in full space) is unambiguously defined and may as well be called T(full space), because the letter T is the universal symbol for kinetic energy, even in classical mechanics. The beauty of an atom is that it shares this unique definition of average kinetic energy with full space because for full space we have

$$\int_{\text{full space}} dV\,\nabla^2\rho = 0$$

As a result, the kinetic energy of an atom is well defined and is typically denoted by $T(\Omega)$. In summary, we can state that:

$$T(\Omega) = G(\Omega) = K(\Omega) \qquad (7.5)$$

How can we accomplish the remaining task of defining the *total* energy of an atom in a molecule? Clearly the total energy is the sum of the kinetic energy and the potential energy, but the latter is unknown. Matters would be considerably simpler if we could somehow obtain the potential energy from the kinetic energy. Can this be done or do we have to calculate the potential energy independently? It turns out that this is not necessary if we use the virial theorem to our advantage.

7.3 The virial theorem

The *virial theorem* originates from the early days of real gas theory and intermolecular forces (see Box 7.2). This theorem expresses a balance between time averaged kinetic and potential energy of a *bound* system. It is an amazingly general and therefore powerful equation since it makes no detailed mechanical assumptions about the system. Soon after the mathematical formalism of quantum mechanics was established, Slater derived the quantum analogue of the classical virial theorem. A series of algebraic manipulations and the Schrödinger equation suffice to obtain the quantum mechanical virial theorem for a bound system of particles (Levine, 1991):

$$2\int d\tau\psi^*\hat{T}\psi = -\int d\tau\psi^*\left(\sum_j \hat{\mathbf{r}}_j \cdot \hat{\mathbf{F}}_j\right)\psi \qquad (7.6)$$

Box 7.2 A closer look

The virial theorem

The roots of the concept of the virial (from Latin for 'forces') lie in the development of thermodynamics in the nineteenth century. When it was realized that intermolecular forces in *real* gases induce deviations in ideal gas behaviour ($pV = kNT$), the quantity $\sum_{\text{all pairs}} \mathbf{r}_{ij} \cdot \mathbf{F}_{ij}$ was introduced, where $\mathbf{F}_{ij}$ is the force between molecules i and j and $\mathbf{r}_{ij}$ is their vector separation. This quantity is called the *virial of the system* and can be shown to be negative if the intermolecular forces are attractive. Since the *real* gas equation is:

$$pV = kNT + \frac{1}{3}\left(\sum_{\text{all pairs}} \mathbf{r}_{ij} \cdot \mathbf{F}_{ij}\right)_{\text{ave}}$$

it follows that the pressure is reduced compared with an ideal gas, as expected. Note that the average is a *time* average. The quantum analogue of the virial, however, is not a time average but the usual quantum expectation value.

One of the simplest derivations of the classical virial theorem differentiates with respect to time the mechanical quantity of *action* A defined by $A = \mathbf{p} \cdot \mathbf{r} = m\mathbf{v} \cdot \mathbf{r}$ ($\mathbf{v}$ and $\mathbf{p}$ are the velocity and momentum of a single particle of mass m). Upon differentiation we obtain:

$$\frac{\mathrm{d}A}{\mathrm{d}t} = \mathbf{F} \cdot \mathbf{r} + 2E_{\mathrm{k}}$$

where E_{k} is the kinetic energy of the particle. A simple calculation shows that the time average of $(\mathrm{d}A/\mathrm{d}t)$ vanishes if the particle moves in a bounded region. This is basically owing to the fact that for a bound particle $\mathbf{r}$ and $\mathbf{v}$ always remain within certain values, i.e. the particle cannot wander off to infinity and its velocity is bound by the speed of light. Consequently we obtain the *virial theorem* for a particle:

$$2(E_{\mathrm{k}})_{\text{ave}} = -(\mathbf{F} \cdot \mathbf{r})_{\text{ave}}$$

where the right side is called the *virial of the particle*. Since a force is in essence a potential energy gradient, the virial is in fact an average potential energy.

where we have taken quantum mechanical averages rather than the classical time averages (consult Box 7.2 to understand equation (7.6)). Note that the sum runs over *all* the particles (both electrons and nuclei) and that the integration is over the *total space* in which the system exists. The right-hand side of equation (7.6) is the potential energy of the system of particles because it is the expectation value of the dot product of the position and force operator. We conclude that the kinetic energy of a system is always related to its potential energy. Thus if we know the kinetic energy we know the potential energy, which is exactly the term we missed in order to obtain the *total* energy of an atom inside a molecule.

To come still closer to our goal we specify the system to be a molecule, i.e. a bound system of nuclei and the electrons. Within the Born–Oppenheimer approximation, equation (7.6) becomes the *molecular virial theorem* shown in equation (7.7). The term on the right-hand side of equation (7.6) falls into two terms, namely the expectation value of the electronic potential energy and a term depending on the forces on the nuclei:

$$2\int d\tau \psi_{el}{}^{*}\hat{T}\psi_{el} = -\int d\tau \psi_{el}{}^{*}\hat{V}_{el}\psi_{el} - \sum_{\alpha} r_{\alpha}\frac{\partial E}{\partial r_{\alpha}} \tag{7.7}$$

where ψ_{el} is the electronic wave function, $\hat{V}_{el}$ is the electron–electron repulsion and electron–nucleus attraction potential energy operator, E is the total energy of the molecule and r_{α} are the nuclear coordinates. Note that for a *(geometrically) optimized* molecule, all forces on the nuclei vanish and the term $\sum_{\alpha} r_{\alpha}\partial E/\partial r_{\alpha}$ in the molecular virial theorem (equation 7.7) vanishes.

The main question, however, still awaits an answer: how do we define a *total* atomic energy? Simply restricting the integrals of the expectation values in equation (7.7) to the atomic subspace rather than full space may look plausible, but can this be justified quantum mechanically? To guarantee that we proceed correctly towards the atomic virial theorem we take a step back and focus on a more general theorem from which the virial theorem can be obtained. This theorem is called the hypervirial theorem.

7.4 The hypervirial theorem

The hypervirial theorem is a wonderfully simple and powerful equation which leads to the virial theorem as one of its special cases. Its formulation invokes commutators which are reviewed in Box 7.3. The hypervirial theorem can be derived (Levine, 1991) from the time-independent Schrödinger equation applying just a few quantum rules and reads:

$$\int_{\substack{\text{full}\\\text{space}}} dV\left\{N\int d\tau' \psi^{*}[\hat{H}, \hat{A}]\psi\right\} = 0 \tag{7.8}$$

where $\hat{H}$ is the Hamiltonian and $\hat{A}$ is a linear, time-independent operator which is *not* necessarily Hermitian (see Box 7.4). It is important to realize, however, that in the derivation of equation (7.8) we have used the fact that $\hat{H}$ is Hermitian *given that the integration extends over full space.*

If the operator $\hat{A}$ is set to $\hat{\mathbf{r}}$ or $\hat{\mathbf{p}}$ we obtain the so-called Ehrenfest relationships (Bader, 1990) which are quantum analogues of elementary laws of classical mechanics. It is more important for our purposes, however, to set

Box 7.3 A closer look

Commutators

Commutators are another abstract concept introduced by Dirac in analogy with the Poisson bracket of classical mechanics. Basically, a commutator expresses the inequality between a product of operators and the same product with the operators appearing in reversed order. An *operator* is a rule that transforms a given function into another function and is denoted by a 'hatted' symbol (circumflex). For example, if $\hat{D}$ is the operator that differentiates a function with respect to x, then $\hat{D}x^2 = 2x$. If $\hat{A}$ and $\hat{B}$ are operators then the *commutator* $[\hat{A}, \hat{B}]$ is defined as $\hat{A}\hat{B} - \hat{B}\hat{A}$. If the commutator of two operators vanishes, the operators are said to *commute*. Commutators form their own algebra which unfortunately we cannot review here. However, two useful commutator rules are

$$[\hat{A}, \hat{B} + \hat{C}] = [\hat{A}, \hat{B}] + [\hat{B}, \hat{C}]$$

and

$$[\hat{A}, \hat{B}\hat{C}] = [\hat{A}, \hat{B}]\hat{C} + \hat{B}[\hat{A}, \hat{C}]$$

which are used in the computation of commutators we need in order to understand the physics of an atom. No derivations can be supplied here and the results are all given for the x component only.

It is straightforward to prove that commutators of different components of position operators vanish (for example $[\hat{x}, \hat{y}] = 0$) as well as commutators of different components of linear momentum operators (for example $[\hat{p}_x, \hat{p}_y] = 0$). Mixed position–momentum commutators involving *different* components such as $[\hat{x}, \hat{p}_y]$ vanish as well, whereas those with *matching* components such as $[\hat{x}, \hat{p}_x]$ are equal to i in atomic units. These simple relationships, commutator rules of the kind given above, and the fact that $\hat{H} = \hat{T} + \hat{V}$ (where $\hat{H}$ is the Hamiltonian, $\hat{T}$ the kinetic energy and $\hat{V}$ the potential energy operator) arm us to derive the three commutators below. Each commutator involves the Hamiltonian and either the position or the momentum operator or their product. The results are:

$$[\hat{H}, \hat{x}] = -i\hat{p}_x$$

$$[\hat{H}, \hat{p}_x] = i\frac{\partial \hat{V}}{\partial x}$$

$$[\hat{H}, \hat{x}\hat{p}_x] = i\left[-\hat{p}_x^2 + \hat{x}\frac{\partial \hat{V}}{\partial x}\right]$$

The latter expression can be generalized to three dimensions using the familiar dot product, so that:

$$\begin{aligned}[\hat{H}, \hat{\mathbf{r}} \cdot \hat{\mathbf{p}}] &= i[-\hat{p}^2 + \hat{\mathbf{r}} \cdot \nabla\hat{V}] \\ &= i[\nabla^2 + \hat{\mathbf{r}} \cdot \nabla\hat{V}]\end{aligned}$$

Box 7.4 A closer look

An important postulate

Typically, textbooks list five postulates of quantum mechanics according to the Copenhagen interpretation. We cannot review all of them in the space of this book. However, there is one postulate of quantum mechanics that deserves our special attention in connection with atomic properties. We first state it and then explain it 'backwards', i.e. we will review only the ideas we need in order to understand this postulate. The postulate runs as follows:

(a) *To every physical observable there corresponds a linear Hermitian operator.*
(b) *To find this operator replace x by $\hat{x}$ and p_x by $-i\partial/\partial x$ in the classical-mechanical expression for this observable.*

Let us take the new elements one by one. A *linear* operator is one that obeys the following property:

$$\hat{D}[cf(x) + dg(x)] = c\hat{D}f(x) + d\hat{D}g(x)$$

where c and d are constants. It is curious that all quantum mechanical operators are linear. For an operator to be *Hermitian*, the following equation must be satisfied:

$$\int \mathrm{d}\tau f^* \hat{A} g = \int \mathrm{d}\tau g (\hat{A} f)^*$$

where f, g are two 'well behaved' functions, the star denotes the complex conjugate and the integration is over full space, which can be of any dimension. In other words, it is possible, for example, to check if a one-dimensional operator is Hermitian by integrating over the real axis from $-\infty$ to $+\infty$. For our AIM purposes the integration will be a triple (volume) integral in 3D space. An *equivalent* definition follows from the one above if both f and g are set to ψ so that

$$\int \mathrm{d}\tau \psi^* \hat{A} \psi = \int \mathrm{d}\tau \psi (\hat{A} \psi)^*$$

The fact that most textbooks give only the first or the second definition of *hermiticity* may be confusing. However, quite remarkably, the first definition also follows from the second, hence both definitions are truly equivalent. Not all operators are Hermitian, for example $\hat{D} = \mathrm{d}/\mathrm{d}x$ is not but $i\hat{D}$ is and so is the kinetic energy operator. We have opted not to use the handy but more abstract 'bracket' notation in this chapter.

We now understand part (a) of the postulate. Part (b) is merely a 'recipe' to transform classical mechanics into quantum mechanics. Dirac used to comment that at the dawn of formal quantum mechanics it was easy to generate important quantum mechanical equations by making these substitutions rather automatically. The Cartesian position coordinate is denoted by x and $\hat{x}$ is the corresponding operator which can, for example, be bracketed between normalized state functions $\int \mathrm{d}\tau \psi^* \hat{x} \psi$ to yield a component of the dipole moment's expectation value. More advanced texts use the symbol q_i ($i = 1, 2, 3$) where $q_1 = x$. The symbol p_x represents a component of the linear-momentum operator.

$\hat{A}$ to $\hat{\mathbf{r}} \cdot \hat{\mathbf{p}}$ so that we retrieve the total-space virial theorem from the hypervirial theorem. Again we ask ourselves if it is correct just to replace the integration of full space by one over atomic subspace. The way forward is to scrutinize the derivation of the total space hypervirial theorem and to detect at which point a certain step is invalid in view of the subspace integration.

If we obtain the correct subspace version of the hypervirial theorem then we could derive a set of quantum mechanical theorems or relationships for finite atoms. The difference between full space and finite subspace is a consequence of the loss of the Hermitian property of $\hat{H}$ over the atomic subspace. Readers interested in the formal expression of this statement should consult the last part of Box 7.1 because in the main text we put emphasis on the logic, the essence, the interpretation and the driving force behind derivations and equations. Consequently we do not reproduce the full derivation here but simply ascertain that we cannot use any hermiticity in the computation of the expectation value of $[\hat{H}, \hat{A}]$ over a subspace. Rather, we are left with a residual term which can be re-expressed using one of Green's identities (Box 7.1). If we set $f = \hat{A}\psi$ and $g = \psi^*$ in this identity and remember that $N\int d\tau'$ is just a prescription to obtain a one-electron property density, then:

$$\int_{\Omega} dV\left\{N\int d\tau' \psi^*[\hat{H}, \hat{A}]\psi\right\} = \frac{1}{2}\iiint_{\Omega} dV\left\{N\int d\tau'[(\hat{A}\psi)\nabla^2\psi^* - \psi^*\nabla^2(\hat{A}\psi)]\right\}$$
$$= \frac{1}{2}\iint_{S} dS\left\{N\int d\tau'[(\hat{A}\psi)\nabla\psi^* - \psi^*\nabla(\hat{A}\psi)]\right\}\cdot \mathbf{n} \neq 0 \tag{7.9}$$

Since $[\hat{H}, \hat{A}]$ is *not* Hermitian over the atomic subspace its corresponding expectation value is not real. To ensure that the subsystem average of an operator is a real number, one takes the mean of the expectation value and its complex conjugate on both sides of equation (7.9). The hypervirial theorem can then be written as follows:

$$\frac{N}{2}\left\{\int_{\Omega} dV \int d\tau'[\psi^*[\hat{H}, \hat{A}]\psi + \psi([\hat{H}, \hat{A}]\psi)^*]\right\}$$
$$= \frac{N}{4}\left\{\iint_{S} dS \int d\tau'[(\hat{A}\psi)\nabla\psi^* - \psi^*\nabla(\hat{A}\psi) + (\hat{A}\psi)^*\nabla\psi - \psi\nabla(\hat{A}\psi)^*]\right\} \tag{7.10}$$

We now reach our final goal if we replace $\hat{A}$ by $\hat{\mathbf{r}} \cdot \hat{\mathbf{p}}$ in the atomic hypervirial theorem. Using the result of Box 7.3 that $[\hat{H}, \hat{\mathbf{r}} \cdot \hat{\mathbf{p}}] = i\{\nabla^2 + \hat{\mathbf{r}} \cdot \nabla\hat{V}\}$, the left-hand side can be rewritten as $(-i)[2K(\Omega) + V_b(\Omega)]$ where we define the 'basin term' $V_b(\Omega)$ as:

$$V_b(\Omega) = \int_{\Omega} dV\left\{N\int d\tau\psi^*(-\hat{\mathbf{r}} \cdot \nabla\hat{V})\psi\right\} \tag{7.11}$$

The fact that $K(\Omega)$ appears in the basin terms can be verified readily using the definition of $K(\mathbf{r})$ given in equation 7.2. The quantity $V_b(\Omega)$ is the integrated average of the virial of the *Ehrenfest force operator*, $\hat{\mathbf{F}}(\mathbf{r}) = -\nabla\hat{V}(\mathbf{r})$, acting on an electron in the atomic basin.

The right-hand side of the hypervirial theorem leads to the 'surface terms' of the atomic virial theorem, obtained by replacing $\hat{A}$ by $\hat{\mathbf{r}} \cdot \hat{\mathbf{p}} = (-i)\hat{\mathbf{r}} \cdot \nabla$:

$$(-i)\frac{N}{4}\left[\iint_S \mathrm{d}S \int \mathrm{d}\tau' [(\hat{\mathbf{r}} \cdot \nabla\psi)\nabla\psi^* - \psi^*\nabla(\hat{\mathbf{r}} \cdot \nabla\psi) + (\hat{\mathbf{r}} \cdot \nabla\psi^*)\nabla\psi \right.$$

$$\left. - \psi\nabla(\hat{\mathbf{r}} \cdot \nabla\psi^*)]\right] = (-i)\left[-\iint_S \mathrm{d}S\ \mathbf{r} \cdot \overleftrightarrow{\sigma} \cdot \mathbf{n} - \frac{1}{4}\iint_S \mathrm{d}S \nabla\rho \cdot \mathbf{n}\right] \qquad (7.12)$$

where we used the relationship $\nabla(\mathbf{r} \cdot \nabla\psi) = \nabla\psi + \mathbf{r} \cdot \nabla\nabla\psi$ and introduced the *stress tensor* $\overleftrightarrow{\sigma}(\mathbf{r})$, which is defined by:

$$\overleftrightarrow{\sigma}(\mathbf{r}) = \frac{1}{4}\left[N\int \mathrm{d}\tau'(-\nabla\psi\nabla\psi^* + \psi^*\nabla\nabla\psi - \nabla\psi^*\nabla\psi + \psi\nabla\nabla\psi^*)\right] \qquad (7.13)$$

The quantum stress tensor is a dyadic (see Box 7.5) which yields a scalar quantity when participating in a double dot product. It was introduced by Schrödinger in relativistic form but here it plays a role in enabling one to think of an atom as a classical object obeying daily life mechanics. If we introduce $V_s(\Omega)$ as the surface virial term, as opposed to the basin virial term $V_b(\Omega)$, and define it as $V_s(\Omega) = \iint_S \mathrm{d}S\ \mathbf{r} \cdot \overleftrightarrow{\sigma} \cdot \mathbf{n}$, then our atomic virial theorem finally reduces to (knowing that $\iint \mathrm{d}S\nabla\rho \cdot \mathbf{n}$ vanishes):

$$\begin{aligned} -2T(\Omega) &= V_b(\Omega) + V_s(\Omega) \\ &= V(\Omega) \end{aligned} \qquad (7.14)$$

where $V(\Omega)$ is the total virial of the atom, and $T(\Omega)$ replaces $K(\Omega)$, which is valid in view of equation (7.5). This equation is remarkably similar to the virial theorem of an optimized molecule.

This result is of utmost importance and puts the atomic energy on a firm and rigorous footing. Historically, the roots of the theory of AIM lie in the observation that the kinetic energy of a molecular fragment bound by a zero-flux surface in $\nabla\rho$ could be transferred from one molecule to another (Bader and Beddall, 1973). The further development of AIM has been driven by the desire to formulate an atomic virial theorem, which renders the atom into an energetically meaningful fragment, obeying its own virial theorem and therefore sometimes called *virial fragment*. Table 7.1 illustrates the calculation of atomic energies of all atoms in methanal.

Substituting $\hat{A}$ for other properties leads to other atomic theorems, such as the atomic force theorem, the atomic torque theorem, the atomic power theorem and others (Bader and Popelier, 1993). Unfortunately we cannot

Box 7.5 A closer look

Vectors, tensors and their products

Although it is tempting to review vectors, tensors and their products extensively, we again focus on what we need immediately to understand atomic properties. Suppose we express the dipole (vector) $\boldsymbol{\mu}$ induced by an electric field $\mathbf{E}$ via the equation $\boldsymbol{\mu} = \boldsymbol{\alpha} \cdot \mathbf{E}$. The polarizability α is an example of a 2D tensor represented by a 3×3 matrix. Note that the component of the electric field along the z-axis can contribute to components of the dipole moment it is orthogonal with, i.e. μ_x and μ_y.

There is a special type of 2D tensor often called *dyadic* because it can be written as a product between a column vector and a row vector, i.e. the dyadic $\mathbf{D} = \mathbf{AB}$ is defined as

$$\mathbf{AB} = \begin{pmatrix} A_x \\ A_y \\ A_z \end{pmatrix} (B_x \;\; B_y \;\; B_z) = \begin{pmatrix} A_xB_x & A_xB_y & A_xB_z \\ A_yB_x & A_yB_y & A_yB_z \\ A_zB_x & A_zB_y & A_zB_z \end{pmatrix}$$

This product is the opposite of the dot product, $\mathbf{B} \cdot \mathbf{A}$, which is the product between a row vector ($\mathbf{B}$) and a column vector ($\mathbf{A}$) and results in a scalar (or number) rather than a matrix. Note that, following well known rules of matrix algebra the product of a dyadic and a vector is another vector. In fact we have already met an example of a dyadic, namely the Hessian where the vector components forming the product displayed above are ($\partial/\partial x$, $\partial/\partial y$, $\partial/\partial z$). The Hessian of ψ for example is then denoted by $\nabla\nabla\psi$ and is a 3×3 matrix containing all possible second derivatives of ψ. An alternative dyadic would be $\nabla\psi\nabla\psi^*$ which contains $(\partial\psi/\partial x)$ $(\partial\psi^*/\partial y)$ as a possible matrix entry. The important quantity called the *stress tensor* is defined entirely via dyadics.

Table 7.1 The energies of the atoms in methanal (in au).

Atom	$E(\Omega)$
C	−37.4332
O	−75.9020
H	−0.6052
Total	**−114.5456**

discuss these atomic theorems in detail but it should be realized that they prove that atoms can be considered as real and rigorously justified objects with their own mechanical properties.

Atoms in molecules are open systems free to exchange charge (and thus matter) and energy with their environment. A challenging programme would be to formulate the thermodynamics of open systems (far from equilibrium) of atoms in molecules. This should be possible since the atomic hypervirial theorem defines quantum analogues of classical quantities such as pressure for an atom (Bader and Austen, 1997).

Subspace quantum mechanics can be put on even firmer ground by using extrema principles based upon the variational principle (Epstein, 1974) using *variational calculus*. This is a mature branch of mathematics, which can be viewed as an extension of ordinary calculus. Variational calculus considers *functionals*, which can be extremized as a function of functions instead of scalars (i.e. numerical variables). Newton's equation ($F = ma$), for example, is obtained by a minimization of the action integral, using the classical principle of least (or stationary) action. It was derived in the 1950s (Schwinger, 1952) that both classical and quantum mechanics can be derived in their entirety from one single principle, the principle of stationary action. It is through this principle that one is able to extend the predictions of quantum mechanics to the properties of a subsystem and thus the AIM integrated properties. Even a concise discussion on this topic is beyond the scope of this book (see Bader, 1988). Basically the most profound justification of the quantum mechanics of an atom can be given only if one goes back to the ultimate source of full space quantum mechanics.

Summary

- The atoms that we recovered from the topology of ρ can also be justified in a rigorous quantum mechanical sense.
- Subspace quantum mechanics is a generalization of full space quantum mechanics in that it introduces extra surface terms. These terms disappear for the topological atom, thereby making it a rigorously defined subspace with fascinating properties.
- Atomic relationships are usually derived using Gauss' divergence theorem, which equates a volume integral to a surface integral over the bounding surface.
- The kinetic energy of an atom is unique and well defined. This is also true for the kinetic energy of the total system.
- The hermiticity of operators does not hold for subspaces. As a result, a real expectation value can be obtained by taking the average of the expectation integral and its complex conjugate.
- The atomic equivalent of the hypervirial theorem contains surface contributions. From this theorem the atomic virial theorem can be derived as well as other theorems.
- It is appropriate to think of an atom as a real object obeying the mechanics of the macroscopic world.
- Variational calculus yields atomic subspace quantum mechanics by going a step back and deriving full space and subspace quantum mechanics within a common framework.

References

Bader R. F. W. (1988). From Schrödinger to atoms in molecules. *Pure and Applied Chemistry*, **60**, 145–155.

Bader R. F. W. (1990). *Atoms in Molecules: A Quantum Theory*. Clarendon: Oxford, p. 141.

Bader R. F. W. and Austen M. A. (1997). Properties of atoms in molecules: atoms under pressure. *Journal of Chemical Physics*, **107**, 4271–4285.

Bader R. F. W. and Beddall P. M. (1973). Virial partitioning of charge distributions and properties of diatomic hydrides. *Journal of the American Chemical Society*, **95**, 305–315.

Bader R. F. W. and Nguyen-Dang T. T. (1981). Quantum theory of atoms in molecules – Dalton revisited. *Advances in Quantum Chemistry*, **14**, 63–123.

Bader R. F. W. and Popelier P. L. A. (1993). Atomic theorems. *International Journal of Quantum Chemistry*, **45**, 189–207.

Epstein S. T. (1974). *The Variation Method in Quantum Chemistry*. London: Academic Press.

Levine I. R. (1991). Quantum Chemistry 4th edn. Englewood Cliffs, NJ: Prentice Hall, chapter 14.

Marsden J. E. and Tromba A. J. (1981). *Vector Calculus* 2nd edn. San Francisco: Freeman.

Schwinger J. (1952). The theory of quantized fields. I. *Physical Review*, **82**, 914–925.

Further reading

Bader R. F. W. (1991). A quantum theory of molecular structure and its applications. *Chemical Reviews*, **91**, 893–928.

Bader R. F. W., Srebrenik S. and Nguyen-Dang T. T. (1978). Subspace quantum dynamics and the quantum action principle. *Journal of Chemical Physics*, **68**, 3680–3691.

Srebrenik S. and Bader R. F. W. (1975). Towards the development of the quantum mechanics of a subspace. *Journal of Chemical Physics*, **63**, 3945–3961.

Srebrenik S., Bader R. F. W. and Nguyen-Dang T. T. (1978). Subspace quantum mechanics and the variational principle. *Journal of Chemical Physics*, **68**, 3667–3679.

Chapter 8

The Laplacian

Sometimes things have to be concentrated in order to make progress

8.1 The Laplacian

The development of the theory of AIM is part of a research programme which aims at a rigorous interpretation of solutions to the Schrödinger equation in terms of chemical concepts. In order to accomplish this ambitious goal AIM takes the electron density as its starting point. We will learn in this chapter that ρ is not the only scalar function that hides a wealth of chemical information. Indeed, there is a derived function, called the *Laplacian of the electron density* and denoted by $\nabla^2\rho$, which contains a wealth of chemical information complementary to that contained in ρ. For example, the Laplacian supports the valence shell electron pair repulsion model, redefines acidity in terms of the balance between kinetic and potential energy densities, leads to a complementarity principle useful in the study of nucleophilic and electrophilic attack and the prediction of van der Waals complexes and finally predicts preferred sites of protonation.

The Laplacian has been defined mathematically in two ways (see Boxes 5.1 and 7.1). A direct and useful definition is that the Laplacian is the sum of the eigenvalues of the Hessian of ρ. The Laplacian of ρ has a topology of its own which is more complicated than that of ρ. Although a study of the *full* topology of $\nabla^2\rho$, including the analogues of attractor basins and IASs has never been carried out, its critical points have been extensively studied. In spite of a seeming simplicity, the Laplacian of ρ has a rich information content which we will systematically unfold in this chapter. In a way, the Laplacian acts as a magnifying glass and enables one to reveal features of ρ that are not readily apparent in ρ itself. The study of $\nabla^2\rho$ fits perfectly with AIM's convention of avoiding arbitrary reference functions. Indeed, we showed in section 2.1 of Chapter 2 that differentiating ρ to obtain $\nabla\rho$, actually corresponds to taking ρ as its own reference, and led to the definition of atoms and bonds via the gradient vector field. Similarly, $\nabla^2\rho$ is constructed from second derivatives of ρ with respect to position, and its topology will in turn reveal the deeper fine structure of ρ.

Understanding the practical meaning of the Laplacian is not easy. Many textbooks simply state the definition and immediately continue with rules and computations involving the ∇^2 operator. In essence the Laplacian measures the local curvature of a function in all its dimensions, of which ρ has three. Box 8.1 invokes the finite difference technique of numerical analysis to help us understand better the deeper meaning of the Laplacian. This technique demonstrates that when the Laplacian of a function is positive the function is *locally depleted.* Conversely, if the Laplacian is negative at a given point then the function is *locally concentrated.* The information in Box 8.1 helps

Box 8.1 A closer look

The meaning of the Laplacian

It is possible to express the second derivative of a function of a single variable x in terms of a finite difference formula. The following formula, which is correct to second order in the finite difference Δx, can be readily found in texts on numerical analysis:

$$f''(x_0) = \left.\frac{d^2 f}{dx^2}\right|_{x=x_0} \approx \frac{2}{(\Delta x)^2}\left(\frac{1}{2}\left\{f(x_0 + \Delta x) + f(x_0 - \Delta x)\right\} - f(x_0)\right)$$

This equation essentially compares the value of a function $f(x_0)$ at a given point x_0 with the average of the function value at points equidistant left and right of the point x_0. Note that the equation remains invariant if Δx is replaced by $-\Delta x$, in other words it does not matter whether Δx is positive or negative. From this equation it can be easily seen that if $f''(x_0) < 0$ then the function value at x_0 will be higher than the average of the function values of its neighbours. In this sense the function is *concentrated at* x_0. On the other hand, if $f''(x_0) > 0$ then the function is lower at x_0 than at its neighbours. An alternative statement is that the function is *depleted at* x_0 if $f''(x_0) > 0$.

This insight can be transferred to the Laplacian. For example, the Laplacian at (x_0, y_0) in two dimensions is given by the following finite difference formula, where $h = \Delta x = \Delta y$:

$$\nabla^2 f(x_0, y_0) = \left(\frac{\partial^2 f}{\partial x^2} + \frac{\partial^2 f}{\partial y^2}\right)_{x_0,\, y_0}$$
$$= \frac{4}{h^2}\left(\frac{1}{4}\left\{f(x_0 + h, y_0) + f(x_0 - h, y_0) + f(x_0, y_0 + h)\right.\right.$$
$$\left.\left. + f(x_0, y_0 - h)\right\} - f(x_0, y_0)\right) + O[h^2]$$

Again the function value at (x_0, y_0) should be compared with the average value of its four immediate neighbours and linked to the sign of $\nabla^2 f$. In three dimensions this principle remains unchanged.

us to understand that when $\nabla^2\rho(\mathbf{r}) < 0$, the value of $\rho(\mathbf{r})$ is greater than the average of its values over an infinitesimal sphere centred on **r**.

It is important not to confuse the actual value of the function with its Laplacian. In particular, the electron density can have a small value but it can nevertheless be locally concentrated. Near the nucleus we can encounter the reversed situation: at some points ρ is very high but locally depleted. Furthermore, it should be realized that the magnitude of $\nabla^2\rho(\mathbf{r})$ determines how strong the effect of electron density concentration or depletion is. It will become clear in the next section that the electron density concentration and depletion are particularly dramatic very close to the nucleus.

8.2 Valence shell charge concentration

Figure 8.1 shows a real and highly relevant example of the Laplacian in which ρ appears to be quite featureless. The electron density of the free neon atom decreases monotonically from a very high value near the nucleus. The corresponding Laplacian, however, is oscillating heavily and reveals an unexpected internal fine structure of the original function, ρ. Since the electron density of a free atom is spherically symmetric, ρ can be fully described in the radial dimension only. The Laplacian of ρ is not defined at the nucleus (i.e. the origin of the plot of Figure 8.1) because of the presence of a cusp. Very near the nucleus (less than 0.1 au) $\nabla^2\rho(\mathbf{r})$ is extremely negative. Moving away from the nucleus the $\nabla^2\rho(\mathbf{r})$ goes through zero at the distance r_1 and climbs to very high positive values. After reaching its maximum the Laplacian decreases until it again reaches zero at r_2. The zone from the origin to r_2 corresponds to the first principal quantum shell ($n = 1$) called the *K* shell. The zeroes in the Laplacian are in fact spherical nodes bounding regions of (electronic) charge depletion or concentration. The next quantum shell, called *L*, appears from r_2 to infinity. Again it consists of a region of charge concentration between r_2 and r_3 and one of charge depletion from r_3 to infinity. Since this shell constitutes the *valence* shell we call the region of negative Laplacian, between r_2 and r_3, the *valence shell charge concentration* (VSCC). The zone with a radius higher than r_3 is called the valence shell charge depletion. In summary, the outer quantum shell is divided into an inner region where $\nabla^2\rho(\mathbf{r}) < 0$ and an outer one where $\nabla^2\rho(\mathbf{r}) > 0$. In the following sections we will focus all our attention on the inner region or the VSCC.

Finally we should mention that the oscillating pattern between negative and positive regions of the Laplacian is expected since the integral of $\nabla^2\rho(\mathbf{r})$ over all space must vanish (see Chapter 7). Also, the first spherical node occurs at approximately a radius $1/Z$ (Z is the atomic number) which amounts to 0.1 au for neon. This small distance is consistent with the spike-like behaviour of the innermost *K* shell (Figure 8.1b).

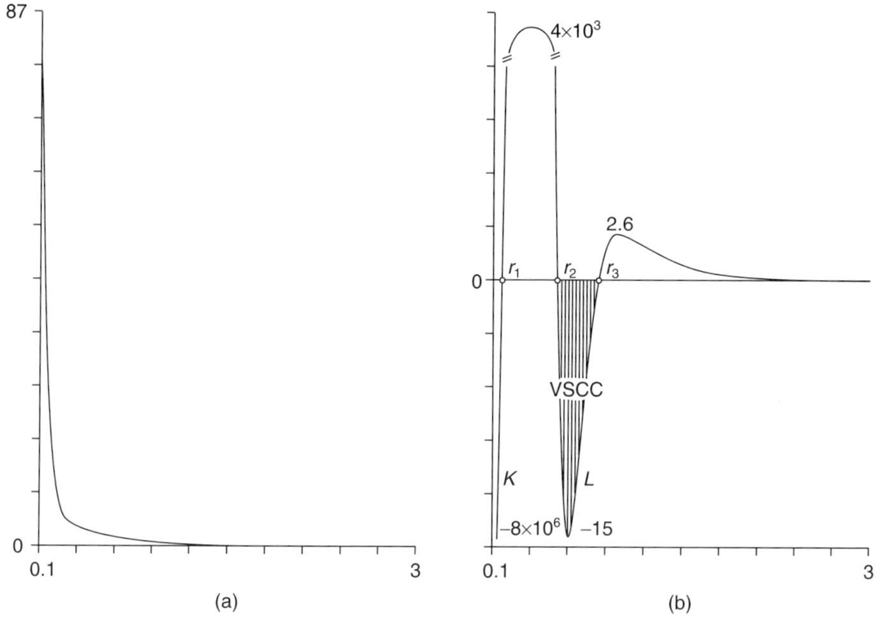

Figure 8.1 (a) The electron density (in au) of a free Ne atom measured along an arbitrary axis (in au) through the nucleus. This function is monotonically decreasing from the origin outward. The part between 0 and 0.1 au has deliberately been omitted because ρ is too high to be represented at the current scale. (b) The Laplacian of ρ (in au) for a free Ne atom computed in the same distance interval as in (a). The Laplacian of ρ is not defined at the nucleus and is extremely negative very near it (-8×10^6 au at 0.1 au). The function $\nabla^2\rho$ passes through zero at r_1 (a spherical node) and climbs to a high value (4×10^3), which cannot be represented at the scale of this plot. The shaded region corresponds to the VSCC (for a full discussion of the profile of $\nabla^2\rho$ see main text).

8.3 The atomic graph

In this section we focus on the topology of the VSCC and show that it extracts chemical information from the wave function that cannot be seen from the electron density. First we introduce the convenient convention of referring to the *negative* of the Laplacian rather than the Laplacian itself. This is because, intuitively, one associates a *positive* value with an electronic charge *concentration* and a *negative* value with a charge *depletion*. For easy reference we introduce a new symbol, $L(\mathbf{r})$, to denote *the negative of the Laplacian* of ρ, or $L(\mathbf{r}) = -\nabla^2\rho(\mathbf{r})$.

Figure 8.2 illustrates this convention by showing a 2D contour plot of $L(\mathbf{r})$ in the symmetry plane of methanal. This diagram demonstrates how the VSCC of the oxygen noticeably deviates from its spherical symmetry

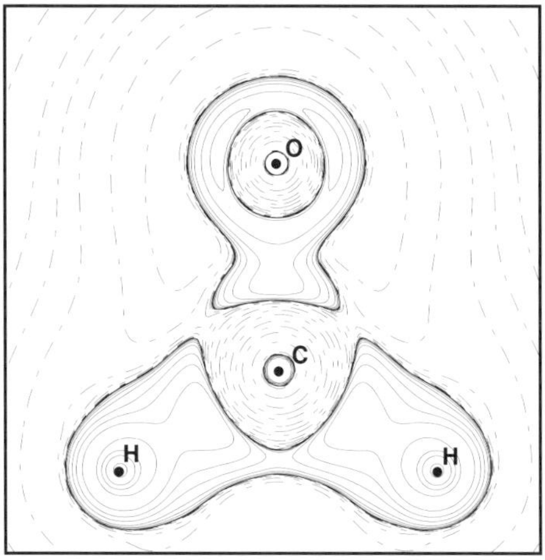

Figure 8.2 A contour map of $L(\mathbf{r}) = -\nabla^2\rho$ in the molecular plane of methanal. The solid contour lines represent positive values of $L(\mathbf{r})$ and the dashed lines negative values of $L(\mathbf{r})$. This function is positive very close to all three nuclei but disconnected only for carbon and oxygen. The hydrogens do not a have a core but share their electrons in the bonds with carbon. For carbon and oxygen, the inner region of electronic charge concentration is surrounded by a region of charge depletion. Both regions together constitute the principal quantum shell *K*, to be compared with the radial profile in Figure 8.1(b). The *L* shell region of charge concentration of oxygen, which is its VSCC, is disconnected from the previously described region of concentration and deviates considerably from spherical symmetry especially at the side of the carbon. Note that in oxygen's VSCC two non-bonding charge concentrations are visible (see further in this section). The VSCC of carbon is severely distorted everywhere and encompasses the two protons.

(as a free atom) upon chemical bonding, especially at the side of the bond with carbon. The principal quantum shells *K* and *L* each still appear as regions of charge concentration and depletion, but the quantum shell *L* is now distorted. The VSCC of the carbon appears to be much more distorted than the VSCC of oxygen for several reasons: it is punctured, contiguous with the VSCC of oxygen and it encompasses both protons.

Figure 8.3(a) shows the full zero envelope of $L(\mathbf{r})$ of methanal, i.e. the set of points where $L(\mathbf{r}) = 0$. Through the punctured VSCC of carbon we can see a small sphere, representing the boundary between the region of charge concentration and depletion of carbon's *K* shell (r_1 in Figure 8.1b). Figure 8.3(b) shows the half of the zero envelope behind the molecular symmetry plane. The *K* shell of oxygen, which is entirely encompassed by its VSCC and therefore hidden in Figure 8.3(a), now becomes visible. Figure 8.3(b) makes clear that the formation of the bond between C and O affects the electron density of the oxygen less than that of the carbon. One should keep in mind that the Laplacian magnifies the small changes induced in the electron

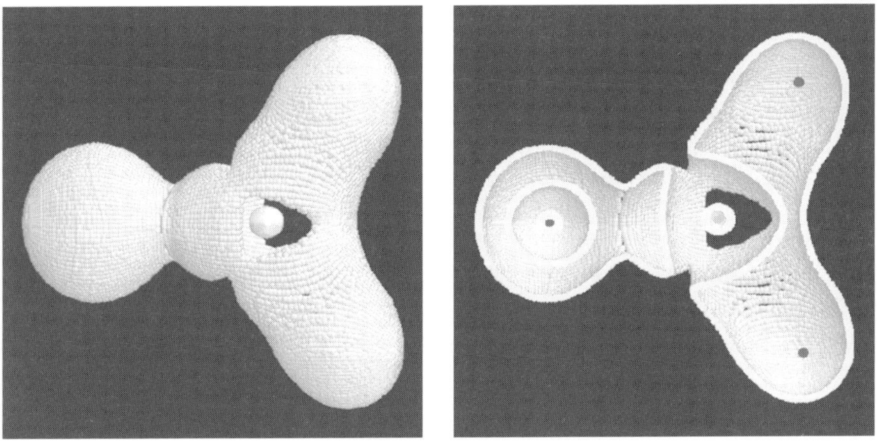

(a) (b)

Figure 8.3 (a) Representation of the zero envelope of $\nabla^2\rho$ of methanal. The small sphere visible through the hole is the surface encompassing the charge concentration of the *K* shell of carbon. (b) A slice through the envelope of part (a) showing the inner zero envelope around oxygen and further details of the shape of the outer envelopes.

density upon bond formation. To quote Bader (1990): 'Just as the energy changes of interest in chemistry are but small fractions of the total energy of a system, so the changes induced in ρ are only small "ripples" in the total density.'

We now focus on the CPs of $L(\mathbf{r})$ and emphasize that the function $L(\mathbf{r})$ has its own topology which should not be confused with that of ρ. Although the same topological concepts appear in one scalar function or another (for example $-\nabla^2\rho$, ρ), the resulting gradient vector fields are *different objects*. The contour diagram of $L(\mathbf{r})$ in Figure 8.2 shows two closed loops within the VSCC of oxygen at either side of the line connecting C and O, near O. This pattern indicates that two maxima or (3, –3) CPs appear in $L(\mathbf{r})$ of the electron density. These CPs are called *non-bonding maxima*. Since they occur in many more cases, such H_2O, it is tempting to identify them as electron density equivalents of the lone pairs in the Lewis model and orbital models. The theoretical basis for this equivalence is obscure but has proven to be extremely useful when supported by a consistent set of observations drawn from many wave functions. The non-bonding maxima of oxygen are shown as well in Figure 8.4 (see Plate section). There is a third (3, –3) CP in the vicinity of oxygen. Since it lies on the bond path between C and O it is called a *bonding maximum*. There is also a bonding maximum in the VSCC of carbon, lying on the bond path between them. Many bonds have two bonding maxima, but sometimes only one maximum occurs, for example in the C–F bond, or none appears as in the P–F bond in PF_3.

Which other types of CP in $L(\mathbf{r})$ can occur in the VSCC of an atom, free or bound? We know from Figure 8.1(b) that near the middle of the VSCC of a free atom, $\nabla^2\rho$ attains a minimum value. Therefore $L(\mathbf{r})$ reaches a maximum value, at a distance from the nucleus which is the radius of a sphere of maximum charge concentration within the VSCC. Consequently, $L(\mathbf{r})$ always possesses at least one negative eigenvalue, corresponding to an eigenvector in the radial direction. Upon chemical binding, this surface of maximum charge concentration persists but becomes distorted and loses its spherical symmetry. The fact that there is always at least one negative curvature of $L(\mathbf{r})$ in the radial dimension remains true for a bonded atom. Thus any CP inside the VSCC will have at least one negative eigenvalue in its spectrum. In other words, a (3, +3) CP can never be encountered in the VSCC since three positive eigenvalues are impossible. In summary, the signature of a VSCC CP can only be −3, −1 or +1; the two types of CP we have not discussed yet are (3, +1) and (3, −1).

Drawing from the previous discussion, how can we interpret a (3, +1) CP? It is *in effect a minimum* on the surface of maximum charge concentration, although it is actually a saddle point in three dimensions. Figure 8.4(b) displays two (3, +1) CPs in $L(\mathbf{r})$ near the oxygen, one above and one below the molecular plane. The profile of $L(\mathbf{r})$ along a line through the oxygen nucleus and a (3, +1) CP shows a maximum in $L(\mathbf{r})$ at the position of the (3, +1) CP, proving that $L(\mathbf{r})$ possesses a maximum in the radial direction. In a plane perpendicular to the radial direction, the function $L(\mathbf{r})$ attains a minimum at the (3, +1) CP. Four (3, +1) CPs appear symmetrically around the carbon nucleus, each near the spot where the VSCC of carbon is punctured. The profile of $L(\mathbf{r})$ along a line from the carbon nucleus through any (3, +1) CP shows that $L(\mathbf{r})$ still attains a maximum in the radial direction, but one for which $L(\mathbf{r}) < 0$. In other words, the maximum appears in a region where $L(\mathbf{r})$ does not rise above zero. As a result it is strictly speaking incorrect to state that the four (3, +1) CPs belong to the VSCC of carbon, but in practice this observation is usually ignored.

Finally, Figure 8.4 shows three (3, −1) CPs in $L(\mathbf{r})$ near the oxygen in methanal. These three saddle points lie in the molecular plane and link the three (3, −3) maxima via gradient paths of $L(\mathbf{r})$. These gradient paths are the mathematical analogues of the bond paths in ρ.

In summary, we discovered three types of CP in $L(\mathbf{r})$ around the oxygen nucleus: (3, −3), (3, −1) and (3, +1). They constitute one topological object which we call the *atomic graph*. A graph is a pictorial representation of a connectivity scheme, rather than a true geometrical object. Figure 8.5(a) shows the atomic graph of oxygen in methanal. The (3, −3) CPs are the vertices, the (3, −1) CPs the edges and the (3, +1) CPs are the faces. It turns out that the number of vertices (V), edges (E) and faces (F) of an atomic graph obey the following relationship:

$$V + E - F = 2 \qquad (8.1)$$

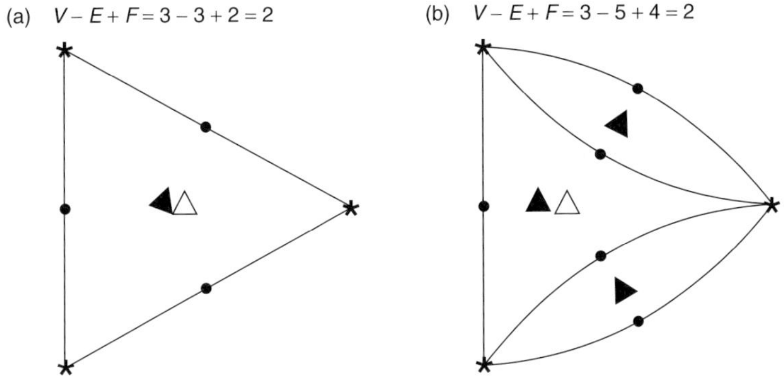

Figure 8.5 (a) The atomic graph of oxygen in methanal. A star represents a (3, −3) CP or vertex (V), a dot a (3, −1) CP or edge (E), and a triangle a (3, +1) CP or face (F). The pair of triangles (solid and open) indicates that the face has two sides (front and back). Alternatively, one face may be thought of as that portion of the plane surrounding the whole planar graph (infinite graph). This graph is planar because it obeys Euler's polyhedral formula ($V - E + F = 2$). (b) The atomic graph of carbon in methanal (same conventions and comments as in (a)).

Euler showed that graphs which obey this relationship are *planar graphs*. This means that the graph can always be drawn on a piece of paper without any edges crossing each other. This property enables a convenient representation and perhaps classification of atomic graphs.

The atomic graph of carbon in methanal is more complicated but can be deduced from Figure 8.4. Figure 8.5(b) shows that, as expected, carbon has only bonding maxima, i.e. three (3, −3) CPs, each one representing a bond. It also has four (3, +1) CPs appearing in a more-or-less tetrahedral arrangement. Note that these (3, +1) CPs occur where the VSCC is punctured. Consequently the VSCC ceases to exist here but the (3, +1) is still a maximum in the radial direction, as explained above.

The correspondence between the number and relative positions of local charge concentrations does not prove the Lewis model, which is based on localized pairs. As discussed in the next section, however, a direct analysis of the wave function (via the pair density) does not support the existence of localized pairs of electrons other than in the atomic core regions. In other words, the presence of a (3, −3) CP in $L(\mathbf{r})$ does not imply the existence of a localized pair of electrons. Nevertheless, there is an abundance of examples demonstrating that there is a mapping between the local maxima in $L(\mathbf{r})$ and the assumed pairs of the Lewis model. This mapping is not always faithful (for example, in very polar molecules such as PF_3) but is more successful than otherwise. One of the great successes of the mapping is that it supports the well known *valence shell electron pair repulsion* (VSEPR) model of Gillespie and Nyholm.

8.4 The VSEPR model

It is not our intention to discuss the VSEPR model in great detail here for it can be found in many textbooks. Rather, we focus on one example, that of ClF_3, and show how the Laplacian acts as a physical basis to support the explanation that the VSEPR model provides.

The VSEPR model is an empirical model, based on observations of hundreds of compounds, mainly inorganic. It formulates a few rules to rationalize preferred (i.e. the most stable) geometries of molecules consisting of a central atom surrounded by up to nine ligands (although for seven or eight ligands there are two or three preferred geometries which cannot be distinguished easily). VSEPR is a natural extension of the localized electron pair model of Lewis and operates under two main *VSEPR assumptions*:

1. The valence electron density is spatially localized into pairs of electrons.
2. The geometrical arrangement of the ligands about a central atom is that which maximizes the interelectron pair separation, both for bonding and non-bonding pairs.

Box 8.2 makes clear that the *physical localization of electronic charge is a direct result of the Pauli exclusion principle.* If electrons form pairs with opposite spin, then this is because of an *indirect* mechanism. Indeed, around each electron there is a 'hole' where same-spin electrons cannot be found. This hole is described via a probability distribution in real space expressing that it becomes increasingly unlikely to find a same-spin electron as one approaches the reference electron. Box 8.2 also makes clear that the *pair density*, which is the electron pair analogue of ρ, does not, in general, define regions of space where electron pairs are localized, except in atomic cores. Nevertheless, there are many examples where the geometry of the CPs in $L(\mathbf{r})$ corroborates the view of the VSEPR model.

Rather than review the three subsidiary postulates of the VSPER model we illustrate the use of the Laplacian in the case of ClF_3. The most stable conformation of this molecule is the T-shaped C_{2v} conformation shown in Figure 8.6, which is the only one observed. That this is the most stable conformation is confirmed both by experiment and by *ab initio* calculations. How can we explain this fact in terms of a simple model? The VSEPR model views this molecule as a trigonal bipyramidal configuration of five electron pairs, arranged on a sphere centred on the central atom. The first correspondence between AIM and VSEPR is that this spherical surface can be associated with the VSCC of the central atom. The second correspondence is that the localized pairs of electrons in VSEPR can be identified with the (3, −3) CPs in $L(\mathbf{r})$. In other words, *the topology of the Laplacian of* ρ *supports the first basic assumption of VSEPR*, stated above, namely that the valence electron density is spatially localized into pairs of electrons.

In addition to the two VSEPR assumptions, the VSEPR model introduces three *VSEPR postulates*. We will now discuss these postulates in some detail. The first postulate is that the *non-bonding pair (n) or lone pairs have larger*

Box 8.2 A closer look

Electron pairing

In 1916, Lewis introduced his concept of the electron pair, an idea that still dominates chemical thinking today. He was so strongly convinced of the existence of coupled pairs of electrons that he was inclined to denounce Coulomb's law for small interelectronic separations, an extreme position he later abandoned. Many decades later we face the question, can we deduce the localization of electrons in pairs from the wave function? More precisely, are there regions in real space in which there is a high probability of finding two electrons of opposite spin?

In analogy with the electron density, which was defined in Box 1.1 as

$$\rho(\mathbf{r}_1) = N \sum_{\text{spins}} \int d\tau_2 \int d\tau_3 \ldots \int d\tau_k \ldots \int d\tau_N \psi^*(\mathbf{Q})\psi(\mathbf{Q})$$

we can define the *pair density* via

$$\rho(\mathbf{r}_1, \mathbf{r}_2) = \frac{N(N-1)}{2} \sum_{\text{spins}} \int d\tau_3 \ldots \int d\tau_k \ldots \int d\tau_N \psi^*(\mathbf{Q})\psi(\mathbf{Q})$$

where the summation runs over the spins of *all N* electrons and $d\tau_k$ represent the three spatial coordinates of the *k*th electron. The integration and summation of the latter formula reduces the $4N$-dimensional wave function to a six-dimensional density function. The pair density $\rho(\mathbf{r}_1, \mathbf{r}_2)$ is given by $N(N-1)/2$ times the probablity density that one electron is at $\mathbf{r}_1$ when the other is at $\mathbf{r}_2$. Its integral over the coordinates of both electrons gives $N(N-1)/2$, the total number of *distinct* pairs of electrons.

To measure how the behaviour of the pair density deviates from a simple product of one-electron densities, we introduce the correlation term $f(\mathbf{r}_1, \mathbf{r}_2)$ which is defined via

$$\rho(\mathbf{r}_1, \mathbf{r}_2) = \frac{1}{2}\rho(\mathbf{r}_1)\rho(\mathbf{r}_2)[1 + f(\mathbf{r}_1, \mathbf{r}_2)]$$

The function $f(\mathbf{r}_1, \mathbf{r}_2)$ gauges both the correlation between electrons with opposite spin, *Coulomb correlation*, and the correlation between electrons with identical spin, *Fermi correlation*. The effect of the Fermi correlation on the pair density is so pronounced that the addition of Coulomb correlation does not change its significant spatial features. One can show that the extent to which sets of electrons may be localized in real space is *solely* determined by the spatial properties of the *Fermi hole*. As an electron moves through space, it carries with it a 'vacuum' or 'shadow' which is a region where no electron with identical spin can occur, called the Fermi hole. Put more precisely, the Fermi hole is a distribution function defined relative to an uncorrelated pair density. It determines the *decrease* in the probability of finding an electron with same spin as some reference electron, relative to a given position of the reference electron.

In Figure 1(a) we show the contour map of the Fermi hole density for a reference electron in the proximity of the carbon nucleus in methane. Because the electron is tightly bound in a deep potential well, its Fermi hole is strongly localized in the immediate vicinity. This means that if the reference electron has an α spin, it causes a high decrease in the probability of finding another α electron near the nucleus. The electrons of β spin are unaffected by this Fermi correlation which is why (α, β) pairs are formed. Thus Fermi correlation does not pair up electrons directly. Rather, the pairs are obtained as a result of all other electrons of both α and β spin being excluded from a given region of space. In summary, α and β electrons are *indirectly* 'pushed together', not because they attract each other, but because α electrons avoid other α electrons, and β electrons avoid other β electrons. To make a curious and loose comparison: an electron pair is like a five-year old boy and girl waltzing together in a ballroom. Most likely they dance together not out of mutual attraction but because girls are told not to dance with girls, and boys not with boys.

We have explained the mechanism of pair formation by means of Figure 1(a). The companion figure (b) reveals that electron pairs are not formed in the first place, when we put the reference electron in the valence region of methane. Clearly, the Fermi correlation contour map is now delocalized which means that the forbidden zones for same spin electrons are much more diffuse. Consequently, α electrons will perceive each other not so much as α electrons but as electrons of any spin. This mean that the electron density is shaped more by the Coulomb repulsion which is weaker than the Pauli repulsion. Thus the pair-formation mechanism cannot develop and localized electron pairs cannot be observed.

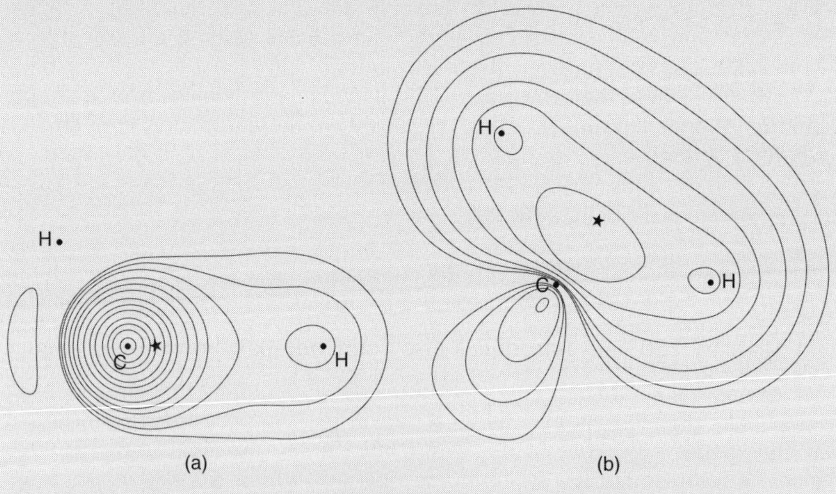

Figure 1 Contour map of the Fermi hole density in the plane containing the carbon nucleus and two hydrogen nuclei in methane. The nuclei lying in this plane are marked by a small dot and the reference electron is marked by a star. In (a) the reference electron, 0.35 au from the C nucleus, is within the localized core of radius 0.53 au. In (b) the reference electron is 0.99 au away from the C nucleus. Figures (a) and (b) represent limiting cases of delocalization, being minimal in the former and maximal in the latter.

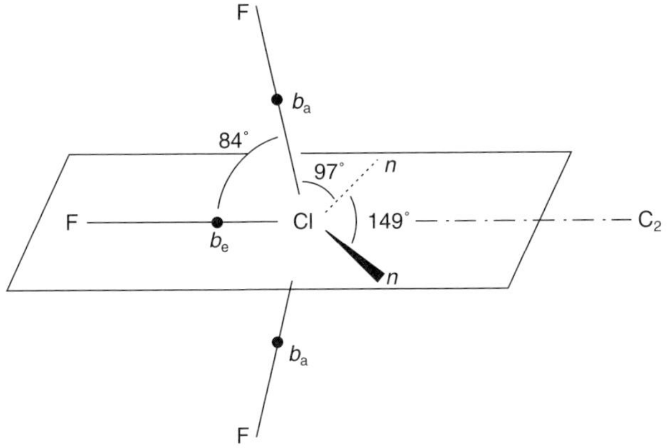

Figure 8.6 Schematic representation of the most stable T-shaped C_{2v} conformation of ClF_3. The bonding (3, −3) CPs in $L(\mathbf{r})$ are marked by b and subscripted by a (axial) or e (equatorial). The non-bonding maxima are marked by n. Note that the axial bonding maximum is actually slightly off the axis connecting Cl and F. The mirror plane, containing F and the two non-bonding maxima is marked, as well as the twofold symmetry axis.

domains than bonding pairs (b) in the same valence shell. Consequently, the angle between two non-bonding pairs is larger than that between a non-bonding and a bonding pair, an angle which is in turn larger than the angle between two bonding pairs. If there is more than one possible location for lone pairs, we choose a ligand arrangement that gives the lone pairs as much room as possible. Let us analyse the molecule ClF_3 (Bader *et al.*, 1984).

Putting the lone pairs in the axial positions (planar conformation, D_{3h}) may well result in their being farthest apart, but is penalized by six lone-pair/bonded-pair contacts of 90° and is therefore ruled out. The arrangement where one lone pair is axial and the other equatorial (tripod conformation, C_s) would result in an undesirable 90° lone-pair contact, further penalized by four lone-pair/bonded-pair contacts of 90°. The remaining possibility where the two lone pairs are equatorial (T-shaped conformation, C_{2v}) leaves them apart by 120° and introduces also four lone-pair/bonded-pair contacts of 90°. Although a rigorous analysis of this molecule according to VSEPR would probably require a little computer program containing a precise penalty function, we correctly conclude here that the T-shaped C_{2v} conformation of ClF_3 is the most stable.

Figure 8.6 schematically represents the geometry of the Laplacian CPs in terms of precise angles between combinations of bonding and non-bonding pairs. The third correspondence between AIM and VSEPR is that the *ranking* of angle values, n–Cl n (149°) > n–Cl–b (97°) > b–Cl–b (84°) is as predicted by VSEPR (these values have been calculated without including electron correlation, see Chapter 1). The experimental value for the F–Cl–F angle is 85°, in excellent agreement with the computed b–Cl–b angle value.

In summary, the topology of $L(\mathbf{r})$ also supports the *second* basic assumption of VSEPR. That is, that the geometrical arrangement of the ligands about a central atom is that which maximizes the interpair separation, given that the electron pairs correspond to the (3, −3) CPs. Moreover, the first VSEPR postulate, stating that the non-bonding pairs take up more space, is also in agreement with the topology of $L(\mathbf{r})$.

A crude estimate of the size of local maxima in $L(\mathbf{r})$ has been proposed in terms of their area on the 'sphere' of charge concentration. This approximate measure, which we will not discuss here (Bader *et al.*, 1984), has provided evidence that the topology of $L(\mathbf{r})$ can support the two remaining postulates of VSEPR as well. In essence, the second VSEPR postulate describes the *change in size of the bonding pair domains as a function of the electronegativities of the ligand and the central atom.* The third VSEPR postulate states that *the domains of double and triple bonds are larger than those of single bonds*, which has been confirmed as well via the area measure. Plenty of molecules illustrating these postulates can be found in the primary literature (for example, Gillespie, 1992).

We conclude that AIM acts as a bridge between the wave function and the VSEPR model. The topology of $L(\mathbf{r})$ constitutes the physical underpinning of a successful empirical model able to predict and rationalize the geometry of many molecules.

8.5 Lewis and the Laplacian complementarity principle

To broaden the Brønsted–Lowry concept of acids and bases, Lewis noted that a Brønsted *acid* was one in which the hydrogen in some molecule (say HCl) *accepted an electron pair* from a second species (say H_2O). Similarly, a *base* (say OH^-) was any chemical species which could *donate an electron pair* to a proton (say H^+ in HCl). Lewis extended the concept of acids and bases to involve the electron pair but did not restrict its acceptance to a proton. He defined a (Lewis) acid as an electron pair acceptor and a (Lewis) base as an electron pair donor. Neutralization is then the formation of a bond (often termed 'coordinate covalent') between an acid and a base. For example, the acid BF_3 and the base NH_3 form the adduct $F_3B{-}NH_3$, which can be regarded as analogous to $H_2O{-}H^+Cl^-$ if the acid is HCl and the base is H_2O. Furthermore, Lewis was able to reinterpret the formation of metal complexes as ligands donating electron pairs to the acid metal ion.

Given that the topology of $L(\mathbf{r})$ parallels the electron pair model of Lewis, which principle involving the Laplacian then parallels the acid–base generalization of Lewis? The answer is simple if we identify an acidic region with one where $L(\mathbf{r}) < 0$ and a basic region with one where $L(\mathbf{r}) > 0$. We then identify a molecule with a local *non-bonding* charge concentration as a *Lewis base* or nucleophile, and one with a local charge depletion as a *Lewis acid* or electrophile. Finally, the 'neutralization principle' of Lewis's

Box 8.3 A closer look

The Laplacian and acidity

One can prove that there is a *local* virial theorem as opposed to the *integrated* virial theorem which we discussed extensively in Chapter 7. In atomic units the local virial theorem is given by:

$$2G(\mathbf{r}) + V(\mathbf{r}) = \frac{1}{4}\nabla^2\rho(\mathbf{r}) \qquad (1)$$

where $G(\mathbf{r})$ is the kinetic energy density defined in equation (7.3) and $V(\mathbf{r})$ is the electronic potential energy density, defined by

$$V(\mathbf{r}) = -\mathbf{r} \cdot \nabla \cdot \overleftrightarrow{\sigma} + \nabla \cdot (\mathbf{r} \cdot \overleftrightarrow{\sigma})$$

where $\overleftrightarrow{\sigma}$ is the stress tensor (dyadic) defined in equation (7.13). Note that if integrated over an atom's basin, equation (1) yields the atomic virial theorem,

$$-2T(\Omega) = V(\Omega)$$

since the integral $\int_\Omega dV \nabla^2\rho$ vanishes.

Equation (1) states in essence that at any point in space there is a balance between kinetic and potential energy density corrected by the Laplacian at that point. This equation yields a novel interpretation of acidity and basicity if we realize that $G(\mathbf{r})$ is everywhere positive and that $V(\mathbf{r})$ is everywhere negative. Indeed, if $\nabla^2\rho > 0$ (or $-\nabla^2\rho < 0$), electronic charge is locally depleted, and it follows from equation (1) that the kinetic energy is *locally* in excess. Thus acidic regions in molecules are in fact zones where, on average, electrons move faster. Conversely, we can conclude that in regions where charge is concentrated ($-\nabla^2\rho > 0$), known as basic regions, the potential energy dominates.

Equation (1) is an expression linking three energy densities. An energy density is dimensionally equivalent to a force per unit area or a pressure. If multiplied by the constant $\hbar^2/4m$, the Laplacian may be viewed as a measure of pressure exerted on the electron density relative to the value of zero.

generalized acid–base theory can be reformulated in terms of the topology of $L(\mathbf{r})$ as follows: *a maximum in* L*(***r***) of an atom in one molecule combines with a minimum (or saddle) in* L*(***r***) of an atom in another molecule.* We call this statement the *Laplacian complementarity principle* and show that it can be applied in a variety of cases. Readers interested in a quantum mechanical interpretation of acidity should consult Box 8.3. Thanks to the existence of the *local* virial theorem we are able to associate acidic regions with zones where electrons move faster on average.

We will now mention a few examples of the complementarity principle expressed via the Laplacian. The first simple and qualitative example is the formation of the well known H_3B–CO adduct. The Lewis base carbon monoxide has a non-bonding charge concentration on the carbon which 'reacts' with the charge depletion (or 'hole') present on the boron atom in BH_3 (see colour picture Figure 7.8 in Bader (1990)).

A second example is the protonation of methanamide, which can be viewed as a reaction where the electrophilic proton attacks the charge concentrations on the oxygen or on the nitrogen atom. The atomic graph of the keto oxygen in methanamide has two large non-bonding (3, –3) CPs in the molecular plane, while nitrogen's atomic graph exhibits two such (3, –3) maxima above and below this plane. The Laplacian correctly predicts the experimentally observed preference of methanamide to be protonated at the keto oxygen. This is so because the absolute value of $L(\mathbf{r})$ is more than twice as high on the oxygen compared with the nitrogen. This example suggests that the complementarity principle has the capacity to make semi-quantitative or at least ranking predictions. The Laplacian has also been used successfully to predict sites of electrophilic attack in substituted benzenes (Bader and Chang, 1989), so its area of application is not restricted to protonation.

A third example considers the (3, +1) CPs appearing in the atomic graph of a keto carbon, above and below the C=O bond axis (see also Figure 8.4b). These CPs form angles of about 110° with respect to the C=O axis, a value which agrees very well with the angle of attack predicted for the approach of a nucleophile to a carbonyl carbon, as measured from crystallographical data (Bürgi and Dunitz, 1983). Revealing sites for nucleophiles to attack is a true asset of the Laplacian-based approach because this is where electrostatic potential maps fail.

A fourth example is the Michael addition reaction of a nucleophile with acrylic acid, CH_2=CH–COOH (Carroll *et al.*, 1989). The Laplacian correctly predicts that the nucleophilic attack occurs at the terminal carbon of the methylene group and that the nucleophile's approach happens from above or below the molecular plane along a line forming an angle of 115° with the C=C bond axis. The latter prediction is in agreement with *independent ab initio* calculations of the potential energy surface of this reaction.

The fifth example focuses on the prediction of geometries of a large number of hydrogen-bonded complexes (Carroll *et al.*, 1988). With only a few exceptions, the Laplacian is able to correctly predict the geometry of a complex of HF with bases such as N_2, CO, CS, CO_2, SCO, HCN, H_2O, HF, O_3, H_2CO and many others. Figure 8.7 illustrates how this is done. A relevant atom in the free base possesses a (3, –3) CP in $L(\mathbf{r})$ which makes a certain angle ($*$–B–X) with the rest of the molecule, denoted by θ_M. The acid HF has a (3, +3) minimum in $L(\mathbf{r})$, located about 0.7 au away from the hydrogen, collinear with the HF bond axis. The (3, –3) CP, the (3, +3) CP and the nucleus B attempt to align, thereby fixing an important degree of freedom in the complex formation. The predicted angle θ_M should then be equal to the angle the minimum (3, +3) makes with the B–X axis, denoted by θ_m, in the case of perfect monomer alignment (see Figure 8.7). Deviations between θ_M

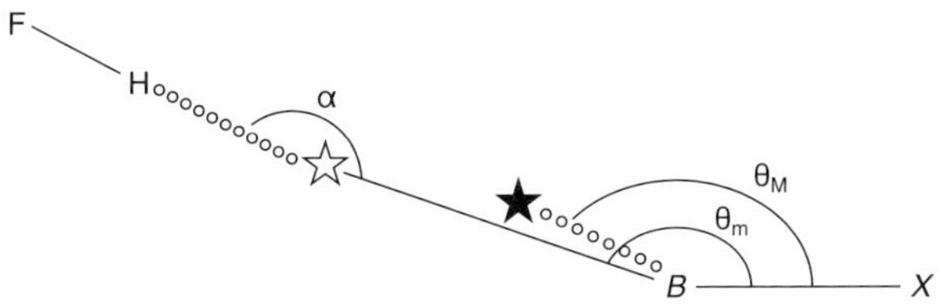

Figure 8.7 Schematic representation of the geometric alignment of HF with a base, using the topology of the Laplacian of ρ. The solid star marks the (3, −3) CP (M) in the vicinity of the base and the open star marks the (3, +3) CP (m) of the acid. A perfect alignment would mean that the angles θ_M and θ_m are identical. Note that the Laplacian-directed alignment does not determine the angle α.

and θ_m measure only a few degrees for several HF-base complexes but can amount to 20 degrees (Carroll *et al.*, 1988). This is still reasonable given the size or extension of the non-bonding zone and the weakness of some hydrogen bond interactions (floppiness of the complexes).

Finally, some research has shown that the Laplacian also recovers the predictions of the frontier orbital model of chemical reactivity. In many cases, the local charge concentration predicted by the Laplacian to be the site of nucleophilic activity coincides with corresponding properties of HOMO (highest occupied molecular orbital) (Bader and MacDougall, 1985). A detailed discussion is beyond the scope of this book.

In summary, the Laplacian is an information-rich function supporting simple models and extracting chemical insight from modern wave functions.

Summary

- The Laplacian of ρ acts as a magnifying glass to reveal chemical features that are not evident in ρ.
- If the Laplacian of ρ is negative at given point then the electron density is locally concentrated. Conversely, if the Laplacian of ρ is positive then ρ is locally depleted.
- The VSCC is the region of electronic charge concentration in the outer quantum shell of an atom. It persists upon chemical bond formation but may be punctured, such as in polar molecules.
- The atomic graph is a connectivity scheme expressing the topology of $-\nabla^2\rho$ in the VSCC of an atom. It consists of three types of CP: (3, −3), (3, −1) and (3, +1).
- The pair density which is rigorously derived from the wave function shows that electron pair localization occurs solely and only to some extent in atomic core regions.

- The mechanism of electron pair formation is indirect and governed mainly by the Pauli exclusion principle. Loosely speaking, a pair of electrons with opposite spin arises as a by-product of the fact that same spin electrons avoid each other and the presence of a nucleus attracting both α and β electrons.
- It is observed that the (3, –3) CPs in the Laplacian can be associated with the electron pairs of the Lewis model, both bonding and nonbonding (or lone pairs).
- The topology of the Laplacian provides a physical basis for the empirical VSEPR model. The Laplacian supports and makes precise the explanations offered by VSEPR.
- The Laplacian principle of complementarity states that a maximum in $L(\mathbf{r})$ of an atom in one molecule combines with a minimum (or saddle) in an atom in $L(\mathbf{r})$ of another molecule. It supports the generalized acid–base principle of Lewis.
- The Laplacian of ρ can be used to study both electrophilic and nucleophilic attack and to predict preferred directions of alignment.
- There is a local version of the virial theorem which enables one to interpret $\nabla^2\rho$ as an energy density or pressure. Acidic regions can be associated with zones of high kinetic energy, and basic regions with high potential energy.

References

Bader R. F. W. (1990). *Atoms in Molecules: A Quantum Theory*. Oxford: Clarendon, chapter 7.

Bader R. F. W. and Chang C. (1989). Properties of atoms in molecules: electrophilic aromatic substitution. *Journal of Physical Chemistry*, **93**, 2946–2956.

Bader R. F. W. and MacDougall P. J. (1985). Toward a theory of chemical reactivity based on the charge density. *Journal of the American Chemical Society*, **107**, 6788–6795.

Bader R. F. W., MacDougall P. J. and Lau C. D. H. (1984). Bonded and nonbonded charge concentrations and their relation to molecular-geometry and reactivity. *Journal of the American Chemical Society*, **106**, 1594–1605.

Bürgi H. B. and Dunitz J. D. (1983). From crystal statics to chemical dynamics. *Accounts of Chemical Research*, **16**, 153–161.

Carroll M. T., Chang C. and Bader R. F. W. (1988). Prediction of structures of hydrogen-bonded complexes using the Laplacian of the charge density. *Molecular Physics*, **63**, 387–405.

Carroll M. T., Cheeseman J. R., Osman R. and Weinstein H. (1989). Nucleophilic addition to activated double bonds: predictions of reactivity from the Laplacian of the charge density. *Journal of Physical Chemistry*, **93**, 5120–5123.

Gillespie R. (1992). Electron densities and the VSEPR model of molecular geometry. *Canadian Journal of Chemistry*, **70**, 642–750.

Further reading

Bader R. F. W., Gillespie R. J. and MacDougall P. J. (1988). A physical basis for the VSEPR model of molecular geometry. *Journal of the American Chemical Society*, **110**, 7329–7336.

Gillespie R. J. (1998). The Physical basis of the VSEPR model. *Structural Chemistry*, **9**, 73–76.

Gillespie R. J. and Hargittai I. (1991). *The VSEPR Model of Molecular Geometry*. Boston: Allyn & Bacon.

Gillespie R. J. and Robinson E. A. (1996). Electron domains and the VSEPR model of molecular geometry. *Angewandte Chemie* (International edition), **35**, 495–514.

Chapter 9

Electrostatics

If your personality had to be summarized by a few words only,
Would not you choose those words carefully?

9.1 The electron population of an atom

In this chapter we discuss a class of atomic properties which are important in two areas: dissecting the internal build-up of a molecule and transferring rigorous atomic information to force fields. These atomic properties are monopole moments (related to atomic charges), dipole moments, quadrupole moments and higher electrostatic moments. As explained in section 3.4, an atomic property is defined as a volume integral of some property density over the basin Ω. In the case of electrostatic moments this property density is the electron density multiplied by a specific geometrical expression, as explained below.

First we focus on the simplest of all electrostatic moments: the *atomic (electronic) monopole moment*, $M_0(\Omega)$, which is defined as follows (using atomic units):

$$M_0(\Omega) = -\int_{\Omega} \mathrm{d}\tau\, \rho(\mathbf{r}) \tag{9.1}$$

In this notation, the letter M refers to the word 'moment' and the subscript zero indicates that this is the zeroth moment. This can be better understood if one realizes that multipole moments actually arise in a series expansion of the electrostatic interaction (Box 9.1). The monopole moment of an electronic charge distribution is trivially related to its *electronic population*, $N(\Omega)$, which is defined as:

$$\begin{aligned} N(\Omega) &= \int_{\Omega} \mathrm{d}\tau\, \rho(\mathbf{r}) \\ &= -M_0(\Omega) \end{aligned} \tag{9.2}$$

In Chapter 3 we computed the population of carbon in methanal to be 5.02 electrons, so carbon's monopole moment is –5.02 au.

Usually, $M_0(\Omega)$ is added to the *nuclear monopole moment*, Z_Ω, which is simply the atomic number of the nucleus. The resulting measure is called the *atomic charge*, $q(\Omega)$, and is defined as:

Box 9.1 A closer look

Multipole moments

Multipole moments arise in the series expansion of a potential which need not be electrostatic but can, for example, be gravitational. Let us reiterate the familiar Taylor expansion in one variable x with respect to a point x_0:

$$f(x) = f(x_0) + f'(x_0)(x - x_0) + f''(x_0)(x - x_0)^2 + \ldots \quad (1)$$

where $f'(x_0)$ and $f''(x_0)$ are the first and second derivative of the function f evaluated at x_0 respectively. If we replace x_0 by the variable y and replace $x - x_0$ by z then equation (1) becomes:

$$f(y + z) = f(y) + f'(y)(z) + f''(y)(z)^2 + \ldots \quad (2)$$

The importance of this equation is that a function depending on two variables y, z is rewritten as a sum of *factorized* terms, i.e. each term is a product of two functions each depending on only *one* variable. Rewriting equation (2) in terms of electron coordinates and nuclear coordinates then leads to a most useful separation of nuclear and electron coordinates, which will become clear below.

Now suppose that we have a molecule centred at a point with position vector **A** and another molecule at **B**. The distance between the two molecular centres is called R, which is the length of the vector $\mathbf{R} = \mathbf{B} - \mathbf{A}$. Each molecule, A and B, possesses an electron density denoted by $\rho_A(\mathbf{r})$ and $\rho_B(\mathbf{r}')$, respectively. The coordinates of both electron densities are deliberately kept different. Figure 1 then clarifies how an electron, with coordinates $\mathbf{A} + \mathbf{r}$ and 'belonging' to $\rho_A(\mathbf{r})$ interacts with another electron at $\mathbf{B} + \mathbf{r}'$, 'belonging' to $\rho_B(\mathbf{r}')$. Omitting the factor $4\pi\varepsilon_0$, the electrostatic energy U_{el} between two electron densities $\rho_A(\mathbf{r})$ and $\rho_B(\mathbf{r}')$, is given by:

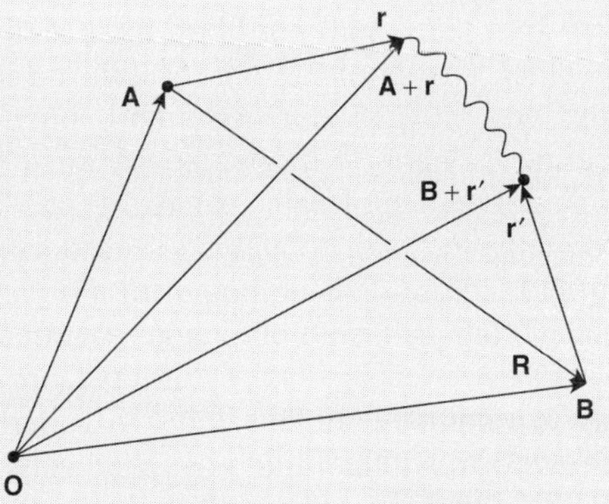

Figure 1 Schematic representation of the vectors **A**, **B**, **r**, **r′**, **R**, **A** + **r** and **B** + **r′** appearing in the derivation of the multipole expansion of the electrostatic energy between two electron densities. Note that $\mathbf{R} = \mathbf{B} - \mathbf{A}$ and that one electron is positioned at **r** and the other at **r′**.

$$U_{el} = \int dr\, dr' \frac{\rho_A(\mathbf{r})\rho_B(\mathbf{r}')}{|(\mathbf{B} + \mathbf{r}') - (\mathbf{A} + \mathbf{r})|}$$

$$= \int dr\, dr' \frac{\rho_A(\mathbf{r})\rho_B(\mathbf{r}')}{|\mathbf{R} - (\mathbf{r} - \mathbf{r}')|} \quad (3)$$

Note that equation (3) is the continuous analogue of the well known expression for the electrostatic energy of a set of discrete charge e_k: $U_{el} = \sum_i \sum_j \frac{e_i e_j}{r_{ij}}$.

The six-dimensional integral in equation (3) cannot be simplified unless we expand $1/|\mathbf{R} - (\mathbf{r} - \mathbf{r}')|$ in a series, along the lines of equation (2). The price we pay for such a series expansion is possible convergence problems, but the enormous benefit is that the electron coordinates $(\mathbf{r} - \mathbf{r}')$ are separated (or factorized) from the molecular coordinates ($\mathbf{R}$). This separation enables us to calculate the multipoles of an electronic charge distribution first, for the isolated molecules A and B, and then compute the interaction between these multipoles in terms of powers of R and angles with vector $\mathbf{R}$. Consequently, we do not have to compute a new wave function of A and B for every new orientation of the two molecules. In other words, the quantum mechanical part of the calculation is completed when the multipole moments are known. After that, it suffices to treat these multipoles classically. Defining the monopole moments M_0 via $M_{0,A} = \int dr \rho_A(r)$ and $M_{0,B} = \int dr' \rho_B(r')$, and the first moments $\mathbf{M}_1$ (or dipole moments μ) via $\mathbf{M}_{1,A} = -\int dr\, \mathbf{r} \rho_A(r)$ and $\mathbf{M}_{1,B} = -\int dr' \mathbf{r}' \rho_B(r')$, leads to the following expression for the electrostatic energy up to first order terms:

$$U_{el} = \frac{M_{0,A} M_{0,B}}{R} + \frac{\mathbf{R} \cdot (\mathbf{M}_{1,B} M_{0,A} - \mathbf{M}_{1,A} M_{0,B})}{R^3} \quad (4)$$

Of course, as they interact the two electron distributions polarize each other and the invariant free-molecule electron density must be modified. Second-order terms in equation (4) would involve quadrupole moments interacting with monopole moments and a dipole moment–dipole moment term, all divided by R^5. It is clear that as R increases the higher-order terms become smaller and U_{el} is well described by the term involving just the monopole moments. The series expansion used to generate equation (4) is convergent only if $|\mathbf{r} - \mathbf{r}'| < R$. Rigorously speaking, the series is formally divergent at any molecular separation, because the molecular electron distributions extend to infinity. However, the series expansion is useful practically except at very short molecular separations (i.e. for small R).

For the derivation of the full equation that equation (4) is part of, we would follow the spherical tensor formalism rather than the Cartesian one. Although there is no difference between these two formalisms if only monopole and dipole moments are included, then dramatic differences in complexity occur with higher moments. A detailed discussion of the spherical tensor formalism is beyond the scope of this work, but we recommend its application, in spite of the use of the non-trivial (but familiar) spherical harmonics rather than just combined powers of x, y and z.

Finally, we must clarify an important point of confusion: the difference between a *multipole* and a multipole *moment*. A *multipole moment* is a term arising in the series expansion of the electrostatic interaction energy between charge distributions. One can compute any multipole moment of any set of charges (discrete or continuous). A *multipole*, on the other hand, is a special arrangement of charges such that all multipole moments except the one corresponding to the multipole vanish. For example, a dipole is a special arrangement of charges such that only its dipole moment does not vanish. Indeed, the monopole moment is zero because the charges of the dipole are equal and opposite, and its quadrupole moment vanishes as well as all higher moments. Similarly, the only surviving moment of a quadrupole is its quadrupole moment.

$$q(\Omega) = M_0(\Omega) + Z_\Omega \tag{9.3}$$

The atomic charge of carbon in methanal is then −5.02 + 6 = +0.98 au. A positive atomic charge indicates that the atom has lost electrons compared with its free, neutral ground state. This atomic property enables one to study rather subtle effects of charge flow such as caused by conformational change. For example, the negative charge on the carbonyl oxygen in a peptide bond is little affected by a rotation around the C–N bond (Milner-White, 1997). In the formation of molecular aggregates, as another example, we observe that the hydrogen attached to the donor atom in a hydrogen bond loses less than 0.1 electron upon hydrogen bond formation (see Chapter 10).

The actual values of atomic charges have been controversial for a long time and the issue about which values are 'correct' is still not resolved today. Undoubtedly the discussion is fuelled by two facts. Firstly there is the lack of an *experimentally* determined atomic charge. Secondly, many atomic charges should actually be regarded more as mathematical entities rather than quantities with a physical meaning. For example, certain numbers, which are called atomic charges, have been obtained by fitting the electrostatic potential to experimental data. When inserted in simple Coulomb-like potentials used in molecular simulations, they may reproduce macroscopic bulk quantities of gases, liquids or solids, rather well, but this is not because they are the actual atomic charges. Indeed, one should not forget that these so-called atomic charges incorrectly absorb several effects owing to polarization, dispersion, higher multipole moments and so forth.

AIM charges, on the other hand, are obtained from quantum mechanics although they do not follow from applying the atomic hypervirial theorem (Chapter 7). Instead, AIM proposes that certain property densities integrated over the atomic basin yield meaningful electrostatic moments. Put more precisely, it is assumed that because the atom has its own well defined energy inside a molecule, its corresponding electronic population is also well defined and meaningful. Perhaps one could argue that the tight relationship between energy and electron density, as expressed by the Hohenberg–Kohn theorem, renders this hypothesis plausible. The AIM electrostatic moments indeed provide a consistent and reliable picture reflecting chemical intuition and thinking. For example, according to AIM, the atomic charge of boron in BH_3NH_3 is +2.0 au, whereas the oft criticised but still widely used Mulliken method yields a charge of –0.3 au. How can a negative charge on boron be compatible with the typical electropositive nature of boron, which releases hydridic hydrogens, acting as a reducing agent?

An exhaustive study of AIM charges of atoms appearing in first row (H) and second row (Li $\rightarrow$ F) diatomics recovers the trends expected on the basis of electronegativity. Figure 9.1(a) shows a sequence of bars representing the charge transfer in seven hydrides *AH*. It is clear from this figure that the hydrogen atom gains electronic charge when bound to an element on its left. The more electropositive this element (*A*), the more negative the charge on the hydrogen. Similarly, more electronegative elements such as N, O and F draw electrons towards themselves, thereby leaving an increasingly positive charge on hydrogen. Note that carbon and hydrogen have an almost identical electronegativity causing charge transfer in hydrocarbons to be rather limited. Based on this link between atomic charges and electronegativity, AIM also concludes that the electronegativity of C relative to H increases with the degree of unsaturation and with the extent of geometrical strain. Figure 9.1 shows similar sequences for the reference atom being carbon or oxygen.

AIM charges have been contrasted with other methods, and although AIM charges are computationally rather expensive they are perceived as reliable (see, for example, Wiberg and Rablen, 1993). The disadvantage of computational expense is steadily dwindling in view of the spectacular advances in computer technology. Since AIM charges are obtained directly from the electron density they depend less on the basis set used compared with methods which separate charge in the (Hilbert) space of the basis functions. Moreover, AIM charges can be computed from experimental electron densities. Nevertheless, AIM charges are sometimes criticized for being too large in absolute value compared with other methods. This observation leads to the incorrect conclusion that the charges would lead to exaggerated dipole moments. We show in the next section that dipole moments based just on charge transfer are wrong.

Finally, it should be mentioned that this extensive discussion perhaps gives the false impression that AIM is just another method of obtaining populations. Although it is unfortunately and wrongly perceived by some to be such a method, AIM was never designed with the sole intent of computing charges. Rather, atomic charges arise as a very useful by-product of a coherent theory rooted in rather different and deeper quantum mechanical considerations.

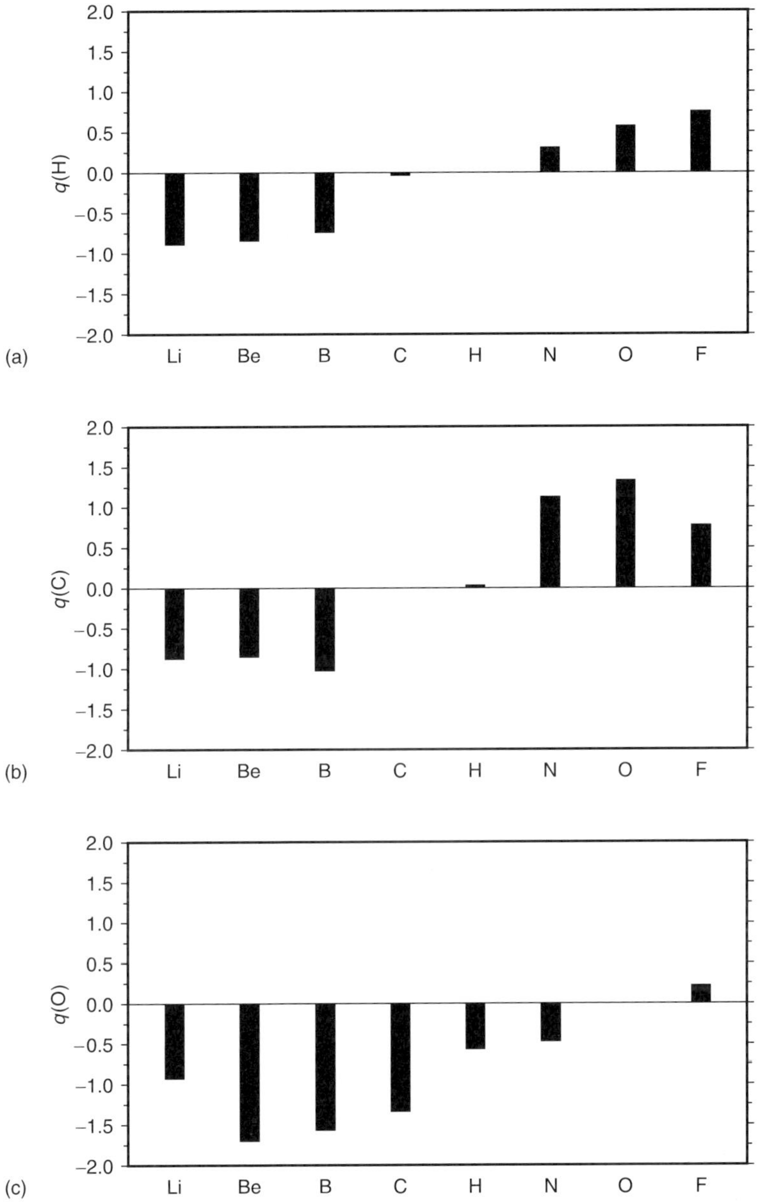

Figure 9.1 Histograms of the atomic charges of A in diatomic molecules AB where B = Li, Be, B, C, H, N, O, F and where (a) A = H, (b) A = C or (c) A = O. All the bars on the right of element A correspond to a positive charge remaining on A as a result of element B drawing electronic population towards itself. This pattern confirms that an element B, appearing in the periodic table right of element A is more electronegative than A. The wave functions are close to the 'Hartree–Fock' limit, i.e. electron correlation has been ignored but large basis sets have been used.

9.2 The atomic dipole moment

The *atomic dipole moment*, $\mathbf{M}_1(\Omega)$, or atomic first moment is a vector consisting of three components. In vector notation it is defined as follows (compare with equation 9.1):

$$\mathbf{M}_1(\Omega) = -\int_{\Omega} d\tau\, \mathbf{r}_{\Omega}\rho(\mathbf{r}) \tag{9.4}$$

where $\mathbf{r}_{\Omega}$ is a vector centred on the nucleus of the atom. The first moment measures the extent, direction and sense of the shift of the atom's electronic charge cloud with respect to the nucleus. The first moment is a vector which points *from* the centroid of the atom's electron density *towards* the nucleus. This perhaps counterintuitive convention of drawing a dipole moment vector is compatible with the rigorous discussions on dipole moments and dipoles in physics textbooks. A dipole (not to be confused with a dipole moment, see Box 9.1) consists of two equal and opposite charges separated by a distance. The electric dipole moment is then defined as a vector going from the *negative* charge to the *positive* charge. Chemists usually draw this vector from the positive charge to the negative charge, which is a convention opposite to the one followed in this text. The chemical convention is inspired by the natural direction of charge transfer but leads to a less straightforward formalism. On the other hand, the physical convention ties in with sign conventions of other closely related concepts such as the potential, the electric field and force.

Figure 9.2(a) illustrates the convention that an atomic dipole moment goes from the centroid of electronic charge (left) to the nucleus (right). This centroid is not a maximum in ρ but is equivalent to the centre of mass in classical mechanics, which is not the point of highest density. Figure 9.2(b) makes clear that an interatomic dipole moment arises from the fact that electronic charge is transferred from the donor atom to the acceptor atom. It is essential to realize that the overall contribution of a bond *DA* to *the total molecular dipole moment always consists of two contributions: a charge transfer term and a polarization term.* For example, it has been shown that the polarization contribution in substituted ethyl and formyl compounds is approximately half the magnitude of the charge transfer term (Slee et *al.*, 1988). Also, in those systems, the atomic dipole moment always opposes the charge transfer dipole moment. The angle between the charge transfer dipole vector and the polarization dipole vector is never lower than 140°. Attempts to assign atomic charges on the basis of measured dipole moments are unrealistic because such procedures ignore the polarizations of the atomic densities.

A prime example of the presence of these two distinct contributions to the molecular dipole moment vector is found in carbon monoxide. The charge on oxygen is −1.15 au. A dipole moment based on just the corresponding charge transfer contribution would clearly lead to a non-vanishing molecular dipole moment. However, it is observed that carbon monoxide has a near-zero dipole moment. This must be owing to considerable atomic po-

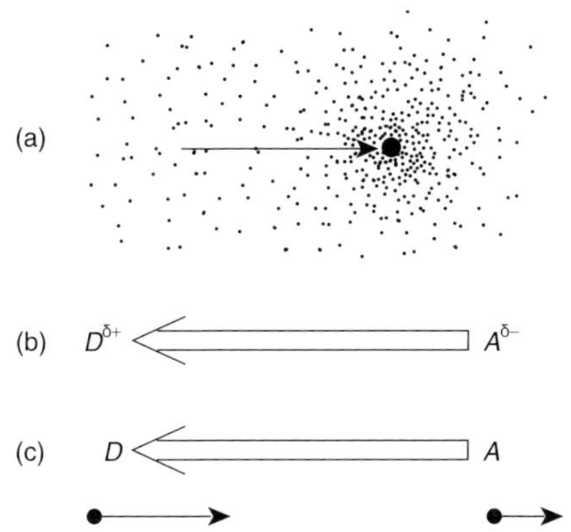

Figure 9.2 (a) Schematic representation of atomic dipole moment which goes from the centroid of electronic charge (left) to the nucleus (right). (b) The interatomic charge transfer dipole moment which arises from two non-zero charges being separated by a given distance. (c) The two contributions (**inter** and **intra**) to the total dipole moment of a bond *DA*. Electronic charge is transferred from the donor atom (*D*) to the acceptor atom (*A*) as measured by the zeroth moments on the atoms. The first moments measure the intra-atomic polarization of the atom or the shift of its electronic charge cloud under the influence of the electric field caused by the charge transfer. Note that the magnitudes of the intra-atomic dipole moments need not be equal. This figure represents a typical situation but is by no means the only pattern for all bonds.

larization effects, counter to the direction of charge transfer. It is indeed found that $|\mathbf{M}_1(\mathrm{C})| = 1.64$ au and $|\mathbf{M}_1(\mathrm{O})| = 0.84$ au, the atomic first moment (i.e. dipolar polarization) vectors both pointing in the direction opposite to the charge transfer. The large dipole moment on carbon is caused by its large, non-bonded and diffusely bound electron distribution. This weakly bound density turns it into a strong Lewis base.

The magnitude of the atomic first moment on oxygen in methanal is 0.44 au and only 0.91 au on carbon, compared with 1.64 au in carbon monoxide. The reduction suggests that the non-bonded density is not present in methanal since carbon is engaged in bonds to the hydrogen atoms. It can be proven that the total molecular dipole moment can be written as follows:

$$\begin{aligned}\mathbf{M}_1 &= \sum_{\Omega} q(\Omega)\mathbf{X}_{\Omega} + \sum_{\Omega} \mathbf{M}(\Omega) \\ &= \mathbf{M}_{\mathrm{CT}} + \mathbf{M}_{\mathrm{atom}} \end{aligned} \tag{9.5}$$

where CT stands for charge transfer, $\mathbf{M}_{\mathrm{atom}}$ is the collection of intra-atomic dipoles (owing to shifts in charge clouds) and $\mathbf{X}_{\Omega}$ are the nuclear positions of the atoms, measured from a common, arbitrary origin. In spite of this arbitrary origin, $\mathbf{M}_{\mathrm{CT}}$ is independent of the choice of origin for neutral molecules, obeying $\sum_{\Omega} q(\Omega) = 0$.

We caution the reader that we have broken down the explanation into two, artificial steps: charge transfer and subsequent polarization. Although this breakdown is supported by rigorous theoretical data, it is not paralleled by a true sequence of effects that can be determined experimentally. Hence this analysis is just a useful mental picture.

9.3 The atomic quadrupole moment

In the previous section we defined the atomic dipole moment which is a vector with three components, explicitly:

$$
\begin{aligned}
M_{1x}(\Omega) &= -\int_\Omega \mathrm{d}\tau\, x\,\rho(\mathbf{r}) \\
M_{1y}(\Omega) &= -\int_\Omega \mathrm{d}\tau\, y\,\rho(\mathbf{r}) \\
M_{1z}(\Omega) &= -\int_\Omega \mathrm{d}\tau\, z\,\rho(\mathbf{r})
\end{aligned}
\tag{9.6}
$$

where x is in fact x_Ω, emphasizing that the vector component is measured from the nucleus, and similarly for y and z. Now the *atomic quadrupole moment* $M_2(\Omega)$ is a vector with five components, explicitly:

$$
\begin{aligned}
M_{2,zz}(\Omega) &= -\int_\Omega \mathrm{d}\tau \frac{1}{2}(3z^2 - r^2)\rho(\mathbf{r}) \\
M_{2,x^2-y^2}(\Omega) &= -\int_\Omega \mathrm{d}\tau \frac{\sqrt{3}}{2}(x^2 - y^2)\rho(\mathbf{r}) \\
M_{2,xy}(\Omega) &= -\int_\Omega \mathrm{d}\tau \sqrt{3}\, xy\rho(\mathbf{r}) \\
M_{2,xz}(\Omega) &= -\int_\Omega \mathrm{d}\tau \sqrt{3}\, xz\rho(\mathbf{r}) \\
M_{2,yz}(\Omega) &= -\int_\Omega \mathrm{d}\tau \sqrt{3}\, yz\rho(\mathbf{r})
\end{aligned}
\tag{9.7}
$$

Each integrand contains an expression in x, y and z such that the sum of all powers of these variables is two in each term. For example, in $M_{2,zz}(\Omega)$, z is raised to the power 2 and in $M_{2,yz}(\Omega)$, y and z are each raised to the power 1, resulting in a total power of 2. This power is the *rank*, ℓ, of a moment. It can be proven that a moment of rank ℓ has $2\ell + 1$ independent components.

Often the quadrupole moment is expressed in terms of six Cartesian components leading to the following expressions appearing in the integrands: xx, xy, xz, yy, yz and zz. Although these integrands are easier to construct than in our definition (equation 9.7), this set of six components has one redundancy. Within the Cartesian formalism, a multipole of rank ℓ generates 3^ℓ components thereby inducing a dramatic increase in redundancies. For this reason, and more advanced reasons beyond the scope of this book, we propose to use the *spherical tensor* formalism to define the components of electrostatic multipoles. This formalism is mathematically more complicated than the Cartesian formalism, and follows a different path to expand the expression $1/|\mathbf{R} - (\mathbf{r} - \mathbf{r}')|$, appearing in the electrostatic interaction (see Box 9.1).

The individual components of any multipole (except the monopole moment which is of rank zero) are dependent on the choice of axis, but their magnitude is not. The *magnitude of a moment* is defined as the square root of the squares of all its components. This definition extends the familiar idea of the magnitude of a 3D vector to an arbitrary dimension. This magnitude is invariant with respect to the choice of the axis system and can therefore be used to compare atoms in different chemical environments without worrying about the molecular orientation. However, an appropriate choice of the axis system may reveal useful information about the details of the distortion in ρ. For example, if the z-axis in water is chosen to be bisecting the H–O–H angle, then the only non-vanishing component of the dipole moment will be the z component. Similar but more complicated considerations can be made for higher multipole moments.

What is the physical meaning of the atomic quadrupole moment? Every moment measures a specific type of distortion of the electron density with respect to an arbitrary axis system centred at the nuclear origin. The quadrupole moment measures how ρ increases or decreases in a direction orthogonal to another one as explained in detail below. Each component of the quadrupole moment measures a particular aspect of this distortion and an appropriate choice of axes can simplify the interpretation of the whole quadrupole moment, as a five-dimensional vector.

For example, Figure 9.3 illustrates schematically the connection between the shape of the electron density and the component $M_{2,zz}$. This component measures how ρ is elongated or compressed along the z-axis relative to a direction perpendicular to z. If this component is negative then the electron distribution is prolate with respect to the z-axis. This fact can be deduced from the definition (equation 9.7), when realizing that ρ weights the (x, y, z) formula that determines the multipole component. Suppose that ρ protrudes along the z-axis. As a result ρ will systematically 'reward' points in space where $3z^2$ is larger than r^2. In other words, the region where $3z^2$ is larger than r^2 (or $3z^2 - r^2 > 0$) contributes more to the integral than the region where $3z^2$ is smaller than r^2 (or $3z^2 - r^2 < 0$). Consequently, $M_{2,zz}$ ends up being negative in view of the minus sign in front of the integral over the atomic volume. Similarly, when the component $M_{2,zz}$ is positive, ρ is oblate. Finally, a spherical electron density results in a value of zero for $M_{2,zz}$. A similar analysis can be performed on the moment M_{2,x^2-y^2} because it measures the shape of ρ with respect to deviations from circular symmetry in the xy

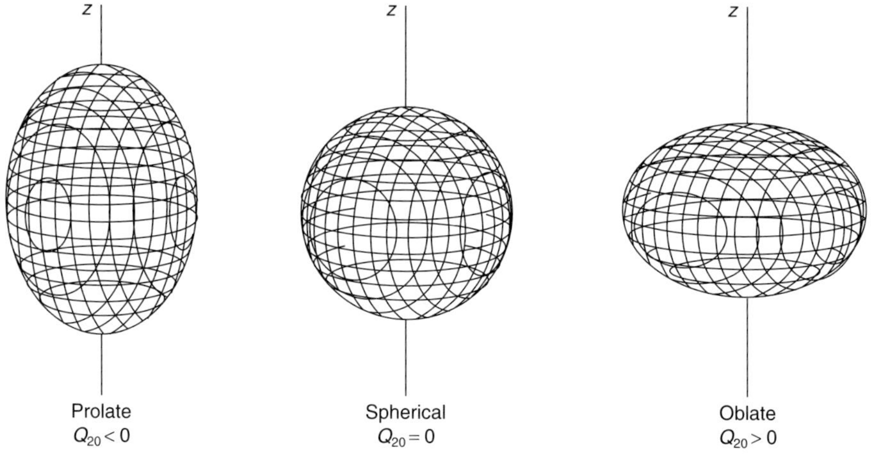

Figure 9.3 Representation of quadrupolar polarizations of an electronic charge distribution. One of the five irreducible quadrupole moments called Q_{20} measures how much the distribution is protruding (when negative) along the z-axis or how much it is flattened (when positive) in that direction.

plane. The analysis of $M_{2,xy}$ involves a comparison of the upper right and lower left quadrants (++ and −−) versus the lower right and upper left quadrants (+− and −+).

With an eye on Figure 9.3 we could wrongly conclude that a prolate ellipsoid is actually the same object as the oblate ellipsoid, in that a different choice of z-axis can transform one into the other. Indeed, the prolate and oblate ellipsoid are intrinsically distinct objects, regardless of their orientation in space. This can be seen when realizing that an ellipsoid is completely determined by its three semi-axes. It is then clear that an ellipsoid with two small semi-axes and one large one can never be superimposed on an ellipsoid with two large semi-axes and one small one.

There are strong indications that the quadrupole moment's magnitude acts as the electron density analogue of a π population of the orbital model. Again, an appropriate choice of axis system, in particular the z-axis, reveals exactly how ρ is distorted with respect to a local axis system. The sequence ethane, benzene, ethene and ethyne contains C–C bonds, formally of bond order 1, 1.5, 2 and 3, and is therefore an ideal test case to measure increased π-character in a C–C bond. The magnitude of the quadrupole moment of carbon, denoted by M_2(C), increases as follows for the sequence given above: 0.10, 1.65, 1.73 and 2.17 au. If we put the z-axis through the carbon nucleus and perpendicular to the molecular plane, we find that the most contributing component is $M_{2,zz}$(C), being −1.64 au for benzene and −1.70 au for ethene. The sign of this component does indeed indicate that ρ is elongated along in the direction orthogonal to the molecular plane, as expected for π systems.

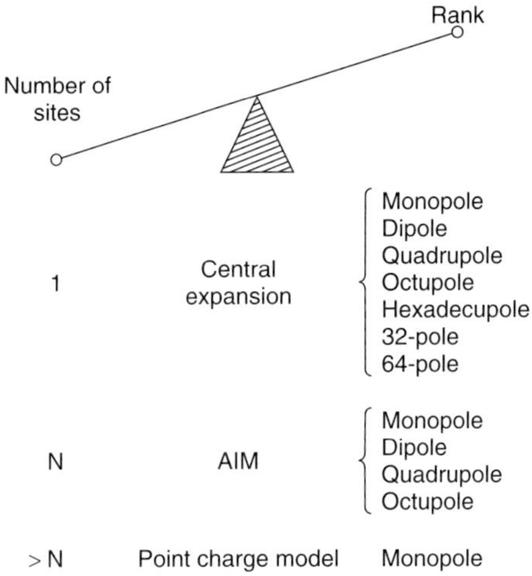

Figure 9.4 Schematic representation of trade-off between a number of multipole-moment sites and the rank of the multipole moments at those sites. The lower the rank of the multipole moments the more are required to generate the electrostatic potential of the charge distribution correctly. AIM multipole moments on each nucleus (N is the total number of nuclei) offer a compromise between the one-site central expansion and typical force-field point charge models.

9.4 The electrostatic potential

The electrostatic potential is a quantity that can be measured, for example by X-ray diffraction. It can also be computed by *ab initio* methods. As explained in Box 9.1, the electrostatic potential caused or generated by a charge distribution can be expanded in terms of electrostatic moments. In principle, the electrostatic potential could be obtained exactly if all moments were included in the series expansion. Reproducing the electrostatic potential from the corresponding moments is an important way of testing the quality of the moments. Such a comparison may show that monopoles or 'point charges' only, centred on the nuclei, are inadequate to represent the electrostatic potential accurately. Nevertheless, extra monopoles can be introduced in order to improve the representation of the potential. Figure 9.4 shows schematically that the electrostatic potential of a molecule can be generated by a spectrum of different multipole expansion schemes. At one end of the spectrum we have the central multipole expansion, which involves only one central site at which all multipole moments are centred. This expansion suffers from convergence problems and requires multipole moments of very high rank (for example 128-pole). At the other end of the spectrum we use only multipoles of rank zero, or point charges, at the expense of having to add many more sites than there are nuclei. The exact positions of these extra sites is a matter of debate. A good compromise is to centre multipole moments of

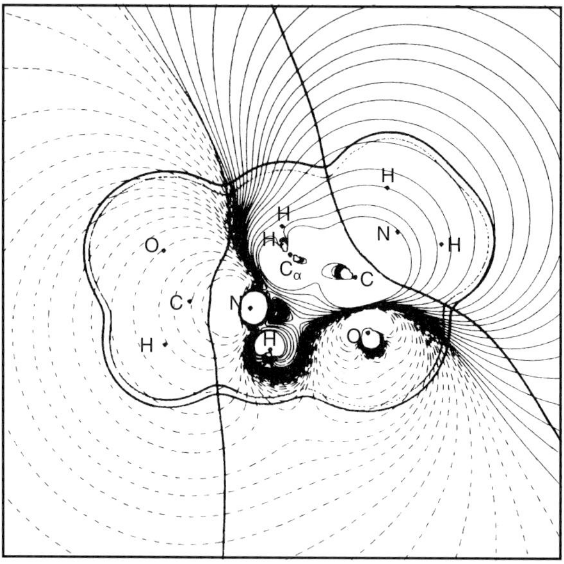

Figure 9.5 Contour map of the electrostatic potential generated by formylglycylamide with values 1, 3, 5, 8, 12, 16, 20, 25, 30, 35, 40, 50, 60, 75, 100, 150, 200, 300 and 400 × 10^{-3} au. Superimposed on this map are the 0.001 au and 0.002 au electron density envelope, inside which the potential is physically meaningless, and two IASs each cleaving the C–N bond of the two peptide bonds. Note that it is possible to transfer the electrostatic potential in between the IASs to a more realistic peptide environment of the glycyl group such as in Gly–Gly–Gly. (From Bader *et al.*, 1992. Reprinted with permission.)

reasonable rank (up to hexadecupole, say) at the nuclei. Figure 9.4 clarifies the trade-off that exists between the number of sites required to generate the electrostatic potential and the number of sites.

Figure 9.5 shows an example of an electrostatic potential using AIM electrostatic moments for a glycyl group, denoted by $|(H)N–C_{\alpha}H_2–C(=O)|$, in a peptide environment. The glycyl group consists of seven atoms, each of which is endowed with multipoles. The formulae used to compute the potential from the multipole moments become rather complex for high ranks but they are simpler than the formulae for the electrostatic interaction. This is because the potential is in fact the energy of a system consisting of a molecule (charge distribution) and a positive unit test charge. Hence, the potential is a special case of the interaction energy and this is reflected in the corresponding formulae. Indeed, equation (4) in Box 9.1, which expresses the interaction energy only in terms of monopole and dipole moments, leads to a simple expression for the potential if we choose B to be the unit test charge. Consequently, $M_{0,B} = 1$ and $\mathbf{M}_{1,B} = \mathbf{0}$, so that we obtain for the electrostatic potential V_{el} of a system A:

$$V_{el} = \frac{M_{0,A}}{R} + \frac{\mathbf{R} \cdot (-\mathbf{M}_{1,A})}{R^3} \tag{9.8}$$

Explicitly, formulae up to R^{-9} are known for the electrostatic interaction (and thus the potential) but will not be given here. All moments up to the hexadecupole moment and interaction energies up to R^{-9} dependence can now be routinely computed within the context of AIM.

Although the potential in Figure 9.5 is plotted inside the van der Waals envelope (represented by the $\rho = 0.001$ au contour) it is only valid outside it. It clearly shows that the system benefits energetically if the test charge is put near the oxygen lone-pair region. In general, the negative contour lines suggest preferred sites of protonation, which is a special case of electrophilic attack. The electrostatic potential approach is not an instrument to study nucleophilic attack because then the 'test object' must be negative. A negative test object carries electrons that interfere with the electron distribution that causes the electrostatic potential, for example via polarization effects. Because of this fundamental complication we cannot rigorously define a classical electrostatic potential any more, in contradistinction to the case when the test charge is just a bare proton. The reader is reminded that the topology of the Laplacian of ρ is able to be used as an instrument to study *both* nucleophilic and electrophilic attack.

It should be mentioned that, as a consequence of the transferability of atoms and their groups, the potential can be transferred as well. This fact enables one to calculate accurate potentials of proteins, drawing from *ab initio* data from oligopeptides (Bader *et al.*, 1992).

Figure 9.6 shows how the electrostatic interaction between two molecules can be treated on a rigorous and general basis. Each molecule is represented

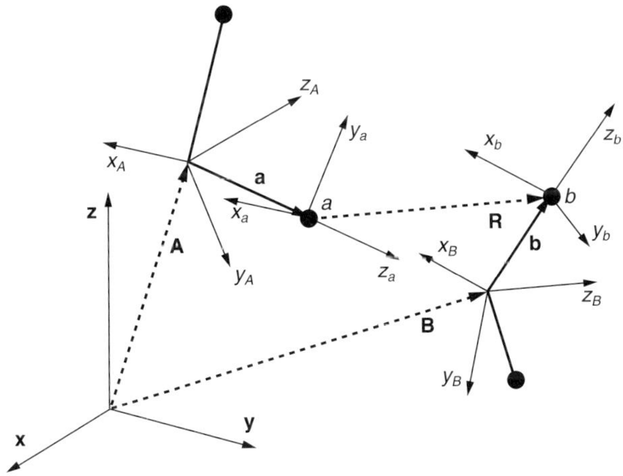

Figure 9.6 Schematic representation of two interacting multipole moments (*a* and *b*) each centred on a different site. Each site is endowed with a local axis system. The full picture illustrates the electrostatic interaction between two molecules each consisting of three atoms (e.g. the water dimer). The molecular axis systems are each centred at **A** and **B**, while the axis system at the bottom left is the global (lab) frame. (From Popelier and Stone, 1994. Reprinted with permission.)

by several sites, each centred on a nucleus, and provided with a local axis system. Each site is endowed with a set of multipole moments expressed with respect to its own axis system (Price *et al.*, 1984). The electrostatic interaction energy between the two molecules *A* and *B* is then written as a sum of site–site interactions. Given that each molecule is treated as a rigid body, the forces and torques induced on the molecule can be obtained and routinely computed for any conglomerate of molecules of arbitrary symmetry (Popelier and Stone, 1994). Moreover, the second derivatives of the interaction energy with respect to rigid body displacement coordinates have been derived such that intermolecular vibration frequencies based on AIM moments can also be calculated.

Finally, distributing the *molecular* electrostatic moments over the constituting nuclei results in an accurate and computationally efficient representation of these molecular moments. These are not the only molecular quantities which profit from such a distribution. Another quantity is the molecular polarizability. Distributing the polarizability over the nuclei leads to distributed atom–atom multipolar polarizabilities, which can be computed *ab initio*. The definition of these distributed polarizabilities based on AIM atomic basins proves to be of great practical benefit (Ángyán *et al.*, 1994).

Summary

- AIM defines an electrostatic moment for an atom inside a molecule as a volume integral of a property density.
- The electronic monopole moment of an atom can be used to study charge transfer in molecules and molecular aggregates (complexes).
- AIM charges have the correct sign but tend to be large compared with other methods. They have proved to be reliable by various assessment criteria. They consistently reflect the chemistry of a wide variety of compounds.
- The total molecular dipole moment always consists of two contributions: a charge transfer term measured by the zeroth moment and a polarization term measured by the first moment. Therefore attempts to assign atomic charges on the basis of dipole moments are unrealistic.
- By convention the dipole moment vector points from the negative charge to the positive.
- The atomic quadrupole moment consists of five independent components. Their values have been proposed as an electron density analogue of a π population of the orbital model.
- The electrostatic potential generated by a (molecular) electron distribution can be expanded using multipole moments. The lower the rank

of a multipole, the higher the number of sites required to represent the potential accurately.

- AIM enables the calculation of the contribution of an atom to the electrostatic potential generated by the molecule it is part of. Since an atom is a transferable piece, its concomitant electrostatic potential is transferable as well. This enables the construction of potential maps of large proteins from accurately computed oligopeptides.

References

Ángyán J. G., Jansen G., Loos M., Hättig C. and Heß B. A. (1994). Distributed polarizabilities using the topological theory of atoms in molecules. *Chemical Physics Letters*, **219**, 267–273.

Bader R. F. W., Popelier P. L. A. and Chang C. (1992). Similarity and complementarity in chemistry. *Theochem-Journal of Molecular Structure*, **87**, 145–171.

Milner-White E. J. (1997). The partial charge of the nitrogen atom in peptide bonds. *Protein Science*, **6**, 2477–2482.

Popelier P. L. A. and Stone A. J. (1994). Formulae for the first and second derivatives of anisotropic potentials with respect to geometrical parameters. *Molecular Physics*, **82**, 411–425.

Price S. L., Stone A. J. and Alderton M. (1984). Explicit formulae for the electronic energy, forces and torques between a pair of molecules of arbitrary symmetry. *Molecular Physics*, **52**, 987–1001.

Slee T., Larouche A. and Bader R. F. W. (1988). Properties of atoms in molecules: dipole moments and substituent effects in ethyl and carbonyl compounds. *Journal of Physical Chemistry*, **92**, 6219–6227.

Wiberg K. B. and Rablen P. R. (1993). Comparison of atomic charges derived via different procedures. *Journal of Computational Chemistry*, **14**, 1504–1518.

Further reading

Stone A. J. (1996). *The Theory of Intermolecular Forces*. Oxford: Oxford University Press.

Chapter 10

Bond characterization

There must be chemistry in all these wave functions
Because we live in one world only.

10.1 Introduction

In Chapter 4 we defined a bond, based on the topology of the electron density. The topology, however, does not specify the character of the bond, it only tells whether it is there or not. Now we are in a position to refine what we mean by a bond using properties evaluated at the BCP. With the aid of these AIM properties we are able to differentiate and classify bonds. Indeed, BCP properties are different for a van der Waals bond and a covalent bond for example.

In this chapter we do not introduce any fundamentally new concepts but rather review previously discussed AIM quantities with the intent of characterizing bonds. Examples of such properties encompass ρ itself, its Laplacian and the ellipticity (ε_b). In this chapter we recover chemical concepts such as bond order, anti-aromaticity, ionicity, (hyper)conjugation and hydrogen bonding from AIM properties. In view of its ubiquity, much attention will be devoted to hydrogen bonds.

10.2 Shared and closed-shell interactions

We remind ourselves of the fact that the Laplacian of ρ at the BCP, $\nabla^2\rho_b$, is the sum of three eigenvalues, λ_1, λ_2 and λ_3. By definition, λ_3 is always positive at the BCP, and measures the curvature of ρ along the AIL. Again by definition, the two other eigenvalues, λ_1 and λ_2, are always negative and measure the curvature of ρ along the IAS, i.e. perpendicular to the AIL. As a consequence the sign of $\nabla^2\rho_b$ determines which curvature dominates in the bonding zone, which is represented by a single point, namely the BCP. However, in the following discussions we will use the quantity $L(\mathbf{r}) = -\nabla^2\rho$ since it is more natural (see Chapter 8). The symbol L_b means the function $L(\mathbf{r})$ evaluated at the BCP.

If $L_b > 0$ then the bond is defined as a *shared interaction*. This name refers to the fact that electronic charge is concentrated in the internuclear region and therefore shared by the two nuclei. The dominant curvatures are λ_1 and λ_2, referring to the contraction of ρ towards the AIL. This type of interaction is expressed in the contiguous region of space where the Laplacian of ρ is negative (see the C–H bond in methanal, Figure 8.2). In view of the local virial theorem,

$$2G(\mathbf{r}) + V(\mathbf{r}) = \frac{1}{4}\nabla^2\rho(\mathbf{r})$$

$$= -\frac{1}{4}L(\mathbf{r})$$

(see Box 8.3), or more conveniently,

$$L(\mathbf{r}) = -8G(\mathbf{r}) - 4V(\mathbf{r})$$

the shared interaction ($L > 0$) is dominated by the lowering of the potential energy $V(\mathbf{r})$ associated with the formation of the BCP, because $V(\mathbf{r})$ is always negative and $G(\mathbf{r})$ always positive.

If $L_b < 0$ then the bond is defined as a *closed-shell interaction*. Here, the dominant curvature is λ_3, stressing that charge is predominantly contracted away from the IAS towards each of the nuclei. In contradistinction to the shared interaction, $L(\mathbf{r})$ is negative at the BCP. Hence, the kinetic energy contribution is greater than the contribution from the potential energy as a consequence of the local virial theorem. The spatial displays of $L(\mathbf{r})$ are atom-like for closed-shell interactions. For a bound state it is the charge, which is concentrated within the basin of each atom, that makes the atoms bind.

Shared and closed-shell interactions are two limiting cases, easily distinguishable and entirely based on the wave function. The question is now, which chemical concepts can be identified with shared and closed-shell interactions? In terms of the above definitions, ionic bonds, hydrogen bonds, bonds in van der Waals molecules (for example Ne–HF) and noble gas clusters (for example Ar_2) are all closed-shell interactions. Moreover, we find that in those systems, the electron density at the BCP, ρ_b, is low, i.e. of the order of 10^{-2} au. On the other hand, covalent or polar bonds are shared interactions, typically leading to ρ_b values of the order 10^{-1} au, roughly an order of magnitude larger than for closed-shell interactions.

Note that on the basis of the present interpretation of AIM quantities it is possible to define and perhaps measure *ionicity* semi-quantitatively. The conditions for an ionic bond are that $L_b < 0$, ρ_b is low (order 10^{-2} au) and that there is considerable charge transfer between the atoms (order 1*e*).

The sequence of second-row diatomic hydrides, *AH*, are perfectly suited to monitor the transition from a typical closed-shell interaction (LiH) to a shared interaction (FH). In every diatomic of this sequence there is a zone of positive L values around the hydrogen nucleus. As the atomic number of the nucleus *A* increases, the BCP gradually moves towards the hydrogen nucleus. In LiH the BCP is in a zone where $L < 0$ and in FH it is in a zone where $L > 0$. In other words, progression through the sequence, LiH, BeH, BH, CH, NH, OH and FH, makes the BCP cross the barrier between the zone of negative and positive *L*. The molecule BeH is transitional in character and said to show an *intermediate interaction*. This interaction typically displays a negative L_b value but also shows a ρ_b value (almost) high enough to typify a shared interaction. The bond between carbon and oxygen in methanal also falls under this category. Finally, we mention that the bond in F_2 is also an intermediate interaction, transitional between N_2 and Ar_2. Indeed, L_b is negative as for Ar_2 but its ρ_b value is about 150 times larger.

A more thorough investigation of the class of intermediate interactions, to which most bonds belong, will lead to a deeper understanding of bond classification. In this section we have explained only a crude classification based on a few properties evaluated at the BCP. Still, precise definitions of ionic and covalent bonds based on *modern ab initio* calculations seem to be lacking at present. The theory of AIM merely shows that the topology of ρ is a promising source of information for the challenging issue of bond classification.

10.3 Bond representation

A bond is a wide concept, encompassing pair-wise attractive interactions ranging from very weak van der Waals bonds such as in Ar_2 to strong covalent bonds such as in N_2. Figure 10.1 shows simple examples of a closed-shell (LiH, Ar_2), an intermediate (CO) and a shared interaction (N_2) via *L* contour maps. In LiH we see clearly that the atoms are separate units, the hydridic hydrogen being polarized towards the Li. The BCP lies well inside the zone where $L < 0$. Essentially the same picture is encountered in Ar_2, but here there is obviously no charge transfer. The picture of CO clearly illustrates the main feature of an intermediate interaction: the proximity of the BCP to the barrier between positive and negative Laplacian.

Recently, Gillespie proposed considering the BF bond in BF_3 as a predominately ionic bond based on AIM charges and a careful deduction of an intramolecular non-bonding radius for the fluorine ligand. His *ligand close-packing (LCP) model* explains the unexpectedly short B–F bond length and makes Pauling's artificial 'back-bonding' correction unnecessary. Based on the fact that $L_b < 0$ and ρ_b is reasonably large (0.22 au), we would classify this bond as intermediate. The LCP model has many more applications. In essence, it proposes that anion-like ligands are close-packed around a cation-like central atom. This model has booked remarkable success in explaining observed bond angle and bond distances as a consequence of this close packing. The development of the LCP model was guided by the use of AIM charges in a series of inorganic molecules. However, a detailed discussion is beyond the scope of this book (Gillespie and Robinson, 1998).

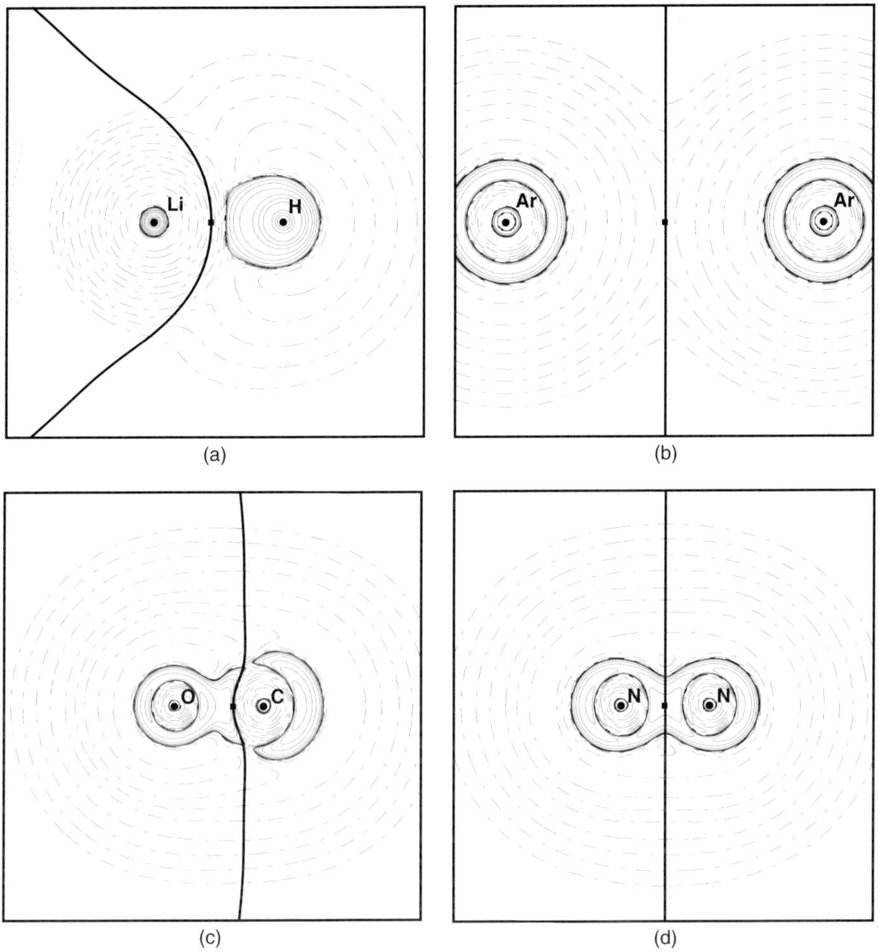

Figure 10.1 The IASs (bold) and BCPs (small square) superimposed on Laplacian contour maps of (a) LiH, (b) Ar_2 (c) CO and (d) N_2. Full lines are contours where $L > 0$, dashed lines where $L < 0$. All pictures are plotted on the same scale.

10.4 BCP properties for a given type of bond

Comparing a given property evaluated at the BCP from one bond to another yields meaningful information provided the nuclei participating in the bond are the same. For example, ρ_b for carbon–carbon (CC) bonds in different chemical environments is related to bond order. In the following discussion we focus on hydrocarbons because they contain a variety of CC bonds which show subtle effects such as hyperconjugation.

Table 10.1 Carbon–carbon bond BCP properties in hydrocarbons.

Molecule	Bond	ρ_b	L_b	ε
Butane	$CH_3CH_2–CH_2CH_3$	0.241	0.510	0.014
Ethene	$CH_2=CH_2$	0.360	1.199	0.298
Ethyne	HC≡CH	0.434	1.464	0.000
Benzene	C_6H_6	0.318	0.937	0.176
Trans-1,3-butadiene	$CH_2CH–CHCH_2$	0.271	0.675	0.071
	$CH_2=CHCHCH_2$	0.358	1.187	0.289
1,3-cyclobutadiene	$\underbrace{CHCH–CHCH}$	0.220	0.394	0.002
	$\underbrace{CHCH=CHCH}$	0.363	1.182	0.353
Propene	$CH_3–CHCH_2$	0.270	0.672	0.029
Toluene	$C_6H_5–CH_3$	0.317	0.929	0.187
Cyclopropane	C_3H_6	0.239	0.419	0.417
Methylcyclopropane	$C_3H_5–CH_3$	0.246	0.541	0.020

Table 10.1 shows ρ_b, L_b and ε_b values of CC bonds in a selection of saturated and unsaturated hydrocarbons. The first three molecules, butane, ethene and ethyne, provide prototypes of a pure single, double and triple CC bond, respectively. It can be seen that ρ_b increases monotonically and that L_b becomes increasingly more positive. These two factors lead to the conclusion that the extent of charge accumulation between the nuclei increases with the bond order, or assumed number of electron pair bonds. A *bond order n* can be defined by successfully fitting the following expression, *A* and *B* being constants:

$$n = \exp[A(\rho_b - B)] \tag{10.1}$$

to about twenty different CC bonds. A typical fit yields a bond order of 1.6 for the CC bond in benzene. This value is intermediate between the Lewis value of 1.50 and the Hückel molecular orbital value of 1.67. The same analytical expression has been fitted with great success to NN, OO, CN or CO bonds. It should be emphasized again, however, that the bond order-ρ_b relationship is distinct for each pair of atoms *A–B*. As a result one *cannot* estimate relative bond strengths for *A–B* bonds from *C–D*, or even *A–C* bonds by comparing their respective ρ_b values.

The ellipticity at the BCP measures the extent to which ρ is elongated in one direction compared to another, both perpendicular to the AIL. In Chapter 5, ε_b was interpreted as an indicator of structural stability in the presence of an RCP. In the case of multiple bonds the ellipticity can acquire a new interpretation, i.e. as a quantitative index of the π-character of the bond. As expected, the ε_b value of the central CC bond in butane is nearly zero, and ε_b increases to 0.176 for the benzene CC bond, and further to 0.298 for the 'pure' double bond in ethene.

Interestingly, this pure double bond ellipticity is somewhat reduced ($\varepsilon_b = 0.289$) for the (formally) double bond in *trans*-1,3-butadiene, while the (formally) single bond in this molecule has an increased ellipticity (0.071 vs 0.014, see Table 10.1). These 'contamination' effects, or deviations from the characteristic ethene value are compatible with the mechanism of *conjugation*. Conjugation is also expressed in the ρ_b value, which is higher for the single bond in *trans*-1,3-butadiene (0.271 au) compared with the pure single bond in butane (0.241 au). Equally, conjugation has the expected effect on the double bonds: $\rho_b = 0.358$ au in *trans*-1,3-butadiene and $\rho_b = 0.360$ au in ethene.

An intriguing question is now, is the extent of conjugation suppressed in an *anti-aromatic* system? Hückel theory predicts cyclobutadiene to be anti-aromatic, and Table 10.1 does indeed show that the single bond ellipticity *decreases* to 0.002 and the double bond ellipticity *increases* to as high as 0.353. Again this phenomenon is paralleled by ρ_b.

The next challenge is to detect *hyperconjugation* in the data in Table 10.1. Orbital models predict that the mechanism of hyperconjugation induces partial double-bond character into what is formally a single bond. Both in toluene and in propene we find a methyl group linked to an unsaturated system. It is clear that the single CC bond displays an enhanced ellipticity and ρ_b value, thereby recovering the anticipated effect of hyperconjugation.

In three-membered rings such as in cyclopropane we observe substantial ellipticities ($\varepsilon_b = 0.417$). This is a consequence of the proximity of the RCP to the BCP (see Chapter 5). Regardless of the difference in origin of this high ellipticity compared with a double bond, we can think of cyclopropyl as being very close in chemical character to a double bond. Chemical similarities between cyclopropane and a double bond have been known for some time, for example in their comparable roles in enhancing the rates of solvolysis reactions. Does this remarkable similarity also mean that a cyclopropyl group leads to hyperconjugation, just like a double bond? The enhanced ellipticity (0.020) of the single C(methyl)–C bond in methylcyclopropane indicates that this is indeed the case. Again, as in propene, partial double-bond character is induced in the bond linking the methyl group to the rest of the molecule.

Finally, we should not over-interpret the ellipticity as a measure for π-character in a bond. In view of the cylindrical symmetry of ethyne, the ellipticity of the triple bond is zero, in spite of considerable π-character. This does not mean that AIM is wrong because ρ is indeed circularly distributed around the bond axis. Rather, it means that we have to be vigilant about the way orbital models are linked with features of the total electron density. In Chapter 9 we proposed the quadrupolar polarization of an atom participating in a CC triple bond as a more reliable quantity to measure π-character. Nevertheless, the ellipticity of the triple bond in molecules such as benzyne or but-1-yn-3-ene deviates slightly from zero, indicating the presence of conjugation.

10.5 Criteria for hydrogen bonding

The electron distribution can be used as a powerful and independent source of information for characterizing hydrogen bonds. Many researchers have more-or-less ignored ρ in the study of hydrogen bonds, lacking a theory that can be used to extract information from ρ. AIM is the ideal tool to detect and classify hydrogen bonds. We will propose eight AIM-based criteria (Koch and Popelier, 1995) which act as a complement to the criteria for hydrogen bonding offered by spectroscopy (IR, NMR), (neutron) diffraction and classical physical chemistry.

One of the main challenges of developing AIM is to observe patterns in the topology and values of scalar fields such as ρ and $\nabla^2\rho$, which coincide with well known chemical concepts such as hydrogen bonding. Ideas of what constitutes a hydrogen bond are in a constant state of flux, however. Consequently, the best we can do is to take a case of hydrogen bonding which is beyond dispute, develop our novel criteria and then apply them in more controversial cases. Many medium to strong typical hydrogen bonds have been studied by AIM and they all comply with the criteria we propose below. Here we introduce, in keeping with the traditional methanal molecule, the methanal–water complex, $H_2CO–H_2O$, which contains one fairly weak O–H · · · O hydrogen bond (about 35 kJ/mol) in a planar conformation shown in Figure 10.2.

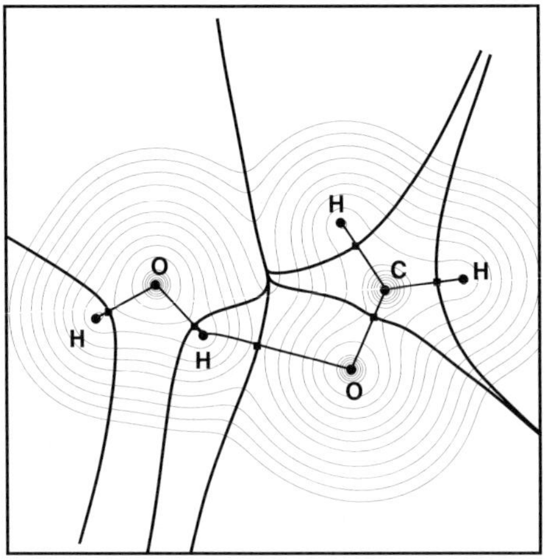

Figure 10.2 The MG and the IASs superimposed on a ρ contour map for a planar conformation of the hydrogen bonded complex methanal–water.

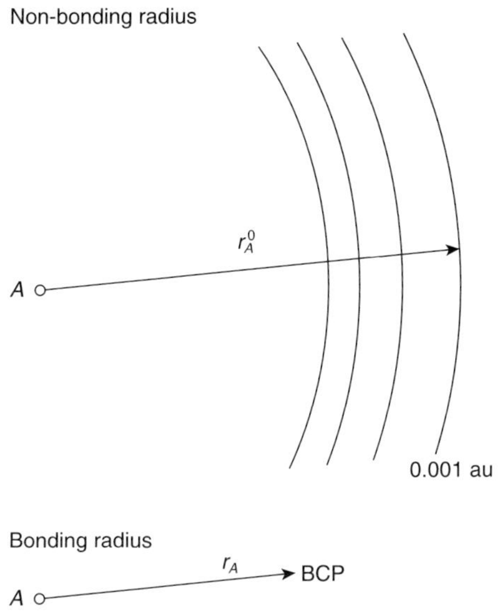

Figure 10.3 A schematic representation of the definition of a non-bonding radius and a bonding radius for an atom *A* participating in a hydrogen bond. The atom *A* is the hydrogen bonding hydrogen or the acceptor atom *B*.

The first criterion is that one has to find a BCP, which topologically proves the existence of a hydrogen bond. Although this may appear to be a trivial requirement, some interactions seem to be expressed merely as geometrical anomalies in observed conformations, rather than via bonds. In Figure 10.2 we do indeed find a consistent topology (BCP, BP, IAS) associated with the hydrogen bond.

Although depending on the level of calculation a meaningful range for ρ_b has been proposed to be [0.002–0.04] au (for neutral monomers), which constitutes the second criterion. In the methanal–water complex we find that $\rho_b = 0.018$ au. It should be noted that an almost perfectly linear relationship between the hydrogen bond energy and ρ_b can be found for a variety of hydrogen-bonded systems, over a wide energy range. The only condition for such a simple relationship is that the chemical environment of the hydrogen bond is kept reasonably constant, e.g. in RCN $\cdots$ HF hydrogen-bonded systems, where *R* may vary from CN to LiO.

As stated before, L_b must be negative for hydrogen bonds, but the third criterion refines this condition by proposing that L_b must lie within the range [−0.15, 0.02] au (for neutral monomers). In the methanal–water complex we observe that $L_b = -0.065$ au.

The fourth criterion demands that the hydrogen (H) and the acceptor atom (*B*) penetrate *each other*. Figure 10.3 makes clear how this penetration can be quantified. The *non-bonding radii* of H and *B*, denoted by r_H^0 and r_B^0, are determined by measuring the distance from the respective nuclei to a given ρ contour in each monomer (or free molecule). Typically the ρ value is taken to be 0.001 au since this is the practical boundary of a molecule. Since this contour is not necessarily symmetrical around the nucleus, one must measure the non-bonding radius in the direction of the hydrogen bond BCP. The *bonding radii*, denoted by r_H and r_B are simply the distances between the respective nuclei and the hydrogen bond BCP. The novelty of AIM is the ability to define bonding radii. The *penetrations* of H and *B*, denoted by Δr_H and Δr_B respectively, are then simply defined via the equations $\Delta r_H = r_H^0 - r_H$, and $\Delta r_B = r_B^0 - r_B$. It has been observed that the hydrogen is penetrated more if it belongs to a soft acidic group, such as C–H, compared with a hard one, such as F–H. In methanal–water, the penetration of both H and O is about 0.8 au.

The following criteria are all invoking atomic properties, which are integrated properties rather than local ones. The fifth criterion states that the hydrogen atom loses electrons, in more precise words its population $N(\Omega)$ decreases. In the methanal–water complex the loss amounts to 0.029*e*. In general, this phenomenon can be related to the descreening of hydrogen bonded protons as observed in NMR chemical shifts.

The sixth criterion demands that the hydrogen atom be *destabilized* in the complex. This effect is measured by $\Delta E(\text{H})$, which is defined as the difference in total atomic energy between the hydrogen in the hydrogen–bonded complex and in the monomer. The value for H in methanal–water is –0.0171 au or −45 kJ/mol.

The seventh criterion asserts that the dipolar polarization (magnitude of the first moment, $\mathbf{M}_1(\Omega)$) of the hydrogen atom *decreases* upon formation of the hydrogen bond. This effect is primarily a result of the loss of non-bonding density of hydrogen in the free monomer. The reduction amounts to 0.022 au in methanal–water.

The eighth and final criterion states that the volume of the hydrogen atom, $v(\text{H})$, should decrease upon complex formation. In methanal–water the reduction is 5.8 au.

It is not clear if all these criteria (summarized in Table 10.2) are actually independent. Strong correlations have been found between the loss of electronic charge and the destabilization of the atomic energy, so perhaps these two criteria can be taken as one. Furthermore, the application of the criteria is hampered for intramolecular hydrogen bonds because most criteria lack the unambiguous reference hydrogen given by the monomer. Nevertheless, they have been successfully applied for those hydrogen bonds under the plausible assumption that a fragment of the molecule can be isolated (via cutting and capping) and used as a monomer-like reference. An example is the bifurcated hydrogen bond in AZT (section 10.6, Figure 10.5 (Plate section)) where the capped fragment 2-methoxypropanol provided appropriate non-bonding radii for the oxygen atoms (Koch and Popelier, 1995).

Finally, it should be mentioned that the agostic bond, which typically arises when a C–H hydrogen binds to a transition metal such as Ti, systematically

Table 10.2 Criteria to detect and characterize hydrogen bonds.

	Criterion
1	Consistent topology (BCP, BP, IAS) for each hydrogen bond
2	A ρ value at the BCP lies within the range [0.002, 0.04] au
3	An $L(\mathbf{r})$ value at the BCP lies within the range [−0.15, −0.02] au
4	Mutual penetration of hydrogen and acceptor atom
5	Loss of charge of the hydrogen atom
6	Energetic destabilization of the hydrogen atom
7	Decrease of dipolar polarization of the hydrogen atom
8	Decrease of the hydrogen atom's volume

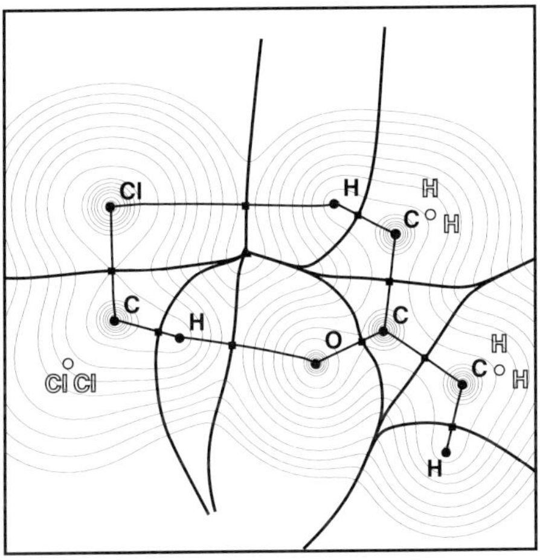

Figure 10.4 Superposition of the MG and IASs on a ρ contour map for the van der Waals complex acetone–chloroform. The (C–)H · · · O bond is a hydrogen bond but the (C–)H · · · Cl bond is not. (From Koch and Popelier, 1995. Copyright © American Chemical Society; reprinted by permission.)

violates all the hydrogen bond criteria (Popelier and Logothetis, 1998). This clearly demonstrates that the assumption that the agostic bond would be a special case of hydrogen bonding is false.

10.6 Cases of hydrogen bonding

In the previous section we asserted eight criteria which are compatible with classical hydrogen bonding. Now we look at an example of a less classical case and see how the failure to comply with a few criteria makes us conclude that the bond is not a hydrogen bond. Figure 10.4 shows a conformation of the acetone-chloroform complex. The (C–)H · · · O bond satisfies all criteria

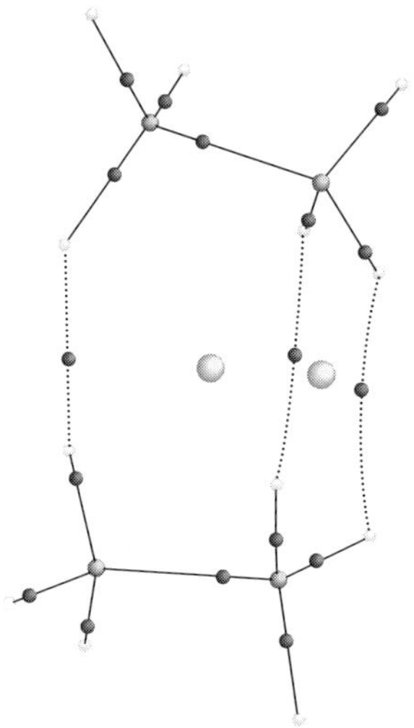

Figure 10.6 Schematic drawing of the MG of the dimer $(BH_3NH_3)_2$ supplemented with the BCPs (black dots) and RCPs (large, light spheres). The dotted lines represent the dihydrogen bonds. (From Popelier, 1998. Copyright © American Chemical Society; reprinted by permission.)

and can therefore safely be called a hydrogen bond. Note that this type of hydrogen bond was controversial until the 1970s. The (C–)H · · · Cl bond, however, fails the penetration criterion because $\Delta r_{Cl} < 0$. Hence, this bond *cannot* be designated as a hydrogen bond.

Figure 10.5 shows an example of an intramolecular hydrogen bond in the anti-AIDS drug AZT. There are two C–H · · · O hydrogen bonds, both satisfying all criteria, of which the C–H · · · O_1 is predicted to be the weaker. This bond will break if the glycosyl torsion angle (see Figure 10.5) is increased. The conformation shown in Figure 10.5 has a zero glycosyl torsion angle, which is unusual. It is tempting to speculate that this unusual conformation, which is stabilized by the bifurcated hydrogen bond, enhances the bioactivity of AZT.

That hydrogen is a creative element in terms of bonding is again confirmed by the *dihydrogen bond* (Richardson *et al.*, 1995). This is a hydrogen bond where the *acceptor* is also a hydrogen atom. Figure 10.6 shows a schematic drawing (with BPs, BCPs and RCPs) of the dimer $(BH_3NH_3)_2$ which contains three dihydrogen bonds. This type of bond again fulfils all hydrogen bonding criteria. As expected, the acceptor hydrogen atom is hydridic which is confirmed by its considerable negative charge of $-0.7e$.

Finally, Figure 10.7 shows two examples of intramolecular hydrogen bonds appearing within peptide structures. The C_5 conformation can contain an unstable N–H · · · O hydrogen bond as part of a five-membered ring. A more stable N–H · · · O hydrogen bond has been observed in a thirteen-membered ring in the 3.6_{13} helical structure of a glycine peptide.

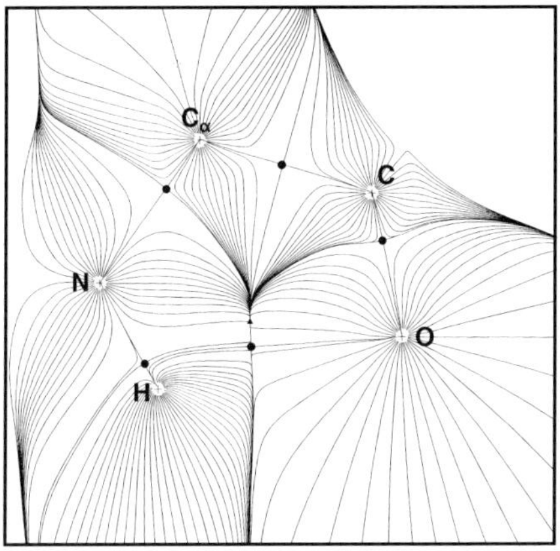

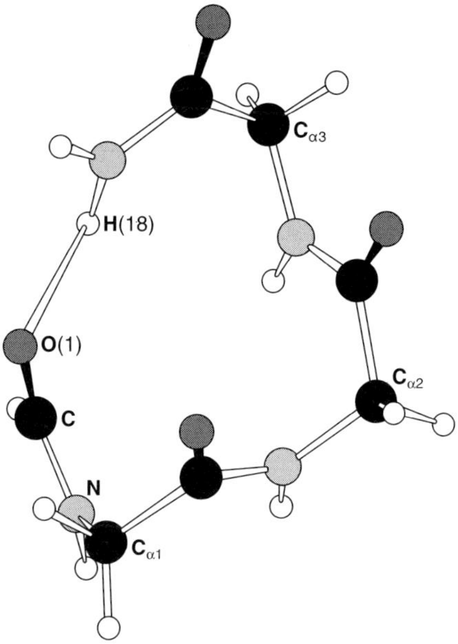

Figure 10.7 (a) The gradient vector field of ρ for the plane of five nuclei in the C_5 structure of *N*-formylglycine amide. The N–H · · · O hydrogen bond is unstable because of the small distance between its BCP and the RCP. (From Bader *et al.*, 1992. Reprinted by permission.) (b) Representation of the 3.6_{13} helical hydrogen-bonded structure of the capped triglycyl peptide HC(=O)–Gly–Gly–Gly–NH_2. The N–H · · · O is part of a thirteen-membered ring. (From Popelier and Bader, 1994. Copyright © American Chemical Society; reprinted by permission.)

Summary

- The topology of ρ determines the presence of a bond. The bond character, however, is not determined by the topology but by properties evaluated at the BCP.
- A shared interaction is one where $\nabla^2\rho_b < 0$, a closed-shell interaction one where $\nabla^2\rho_b > 0$ and an intermediate one where the BCP lies very close to the surface where $\nabla^2\rho_b = 0$. The latter typically combines a positive $\nabla^2\rho_b$ value with a reasonably high ρ_b.
- Rigorous information from the wave function should be transferred to simple bonding diagrams in order to diversify and label bonds in a reliable and compact way.
- It is possible to define a bond order empirically, based on ρ_b, provided the nuclei participating in the bond are the same.

- BCP properties reveal conjugation, hyperconjugation and anti-aromaticity in a wide variety of hydrocarbons.
- AIM allows the study of hydrogen bonds independently from IR, NMR, neutron diffraction and classical physicochemical techniques.
- Eight criteria based on AIM properties have been proposed to characterize both inter- and intramolecular hydrogen bonds. These criteria are well anchored in classical hydrogen bonds and can be used to study more controversial cases.
- Candidate hydrogen bonds which fail one or more of the eight criteria can therefore *not* be designated as hydrogen bonds.
- The criteria have been successfully applied in the dihydrogen bond and biomolecules such as AZT and glycine peptides.

References

Gillespie R. J. and Robinson E. A. (1998). Molecular geometry of 'ionic' molecules: a ligand-close-packing model. *Advances in Molecular Structure Research*, **4**, 1–41.

Koch U. and Popelier P. L. A. (1995). Characterization of C–H–O hydrogen bonds on the basis of the charge density. *Journal of Physical Chemistry*, **99**, 9747–9754.

Popelier P. L. A. (1998). Characterization of a dihydrogen bond on the basis of the electron density. *Journal of Physical Chemistry*, **102**, 1873–1878.

Popelier P. L. A. and Bader R. F. W. (1994) Effect of twisting a polypeptide on its geometry and electron distribution. *Journal of Physical Chemistry*, **98**, 4473–4481.

Popelier P. L. A. and Logothetis G. (1998). Characterization of an agostic bond on the basis of the electron density. *Journal of Organometallic Chemistry*, **555**, 101–111.

Richardson T. B., de Gala S. and Crabtree R. H. (1995). Unconventional hydrogen bonds: intermolecular B–H · · · H–N interactions. *Journal of the American Chemical Society*, **117**, 12875–12876.

Compound index

Subject index

A

B

C

V

W

X

Z